Taking Off Domestic Building Construction

This book provides a detailed step-by-step guide to taking off building work. It is pitched at a basic introductory level especially suitable for technicians entering the construction industry from school, perhaps intending to follow a trade or technical career. An ideal workbook for students of quantity surveying, construction and civil engineering when learning to measure building work. It introduces students to the most basic aspects of measurement and prepares them for learning the more complex areas of taking off.

The book works through examples based on the measurement of a simple traditional pair of semi-detached (duplex) houses, with the relevant construction technology explained throughout. Although the format largely follows United Kingdom taking-off practice and conventions, it will be directly applicable to international practice in countries broadly following United Kingdom conventions. Each chapter presents a worked example from the substructure, through to masonry walls, upper floors, roofing, internal finishes, windows and doors. The examples are matched to an exercise for students and include a commentary of why and how the take-off work is being done. Concentrating specifically on the skilled task of taking off, the examples are designed to give confidence and practice rather than theoretical knowledge. This practical book is ideal for use on the Design, Surveying and Planning T Level; HNC Construction; and undergraduate and non-cognate postgraduate courses in Quantity Surveying, Construction and Building Surveying.

Andy Atkinson is a chartered quantity surveyor with a background in consultancy and public service. With 35 years' experience lecturing in measurement of building work and contract administration at London South Bank University, he has acted as principal and co-investigator for several publicly funded research projects. Formerly a member of the Joint Contracts Tribunal BIM Working Group, examining methods for adapting JCT contracts to building information modelling, Andy is the author of the *JCT Contract Administration Pocket Book* (Routledge 2020). He also maintains a small surveying consultancy.

Taking Off Domestic Building Construction

An Introduction to Building Quantities

Dr Andy Atkinson

PhD, MSc, FRICS, Cert Ed.

LONDON AND NEW YORK

Cover image design: Peter McWilliam and Matthew T Young of MTY Architects, 25 Hunts Mill, Goldsmiths Lane, Wallingford, Oxfordshire, OX10 0DN, UK, based on author's drawing produced using MicroStation® CAD software and Sketchup Go visualisation program.

First published 2024
by Routledge
4 Park Square, Milton Park, Abingdon, Oxon OX14 4RN

and by Routledge
605 Third Avenue, New York, NY 10158

Routledge is an imprint of the Taylor & Francis Group, an informa business

British Library Cataloguing-in-Publication Data
A catalogue record for this book is available from the British Library

ISBN: 978-1-032-18152-3 (hbk)
ISBN: 978-1-032-17160-9 (pbk)
ISBN: 978-1-003-25312-9 (ebk)

DOI: 12.01/9781003253129

Typeset in Times New Roman
by Apex CoVantage, LLC

Printed in the UK by Severn, Gloucester on responsibly sourced paper

Contents

Acknowledgements

Learning, doing and teaching

I learned taking off as a trainee at Henry Cooper and Sons of Reading, reputedly the first consulting quantity surveying practice in the world, working in yards, feet and inches and pounds, shillings and pence without a calculator. Numerous site visits helped develop the visualisation skills necessary to measure effectively, and the mental arithmetic involved embedded in me a fertile source of anecdotes of how things were done back in the day. It is to my mentors at Coopers that I owe first acknowledgements. I quickly moved on to higher education at Thames Polytechnic, where lectures in measurement confirmed and improved my developing skills, and I acknowledge both the tutors and fellow students there for this.

However, it is doing the job in the final analysis that really shows if you are competent. As I often point out to my own students, it is not the certificate of competence that shows you are capable but whether you can get professional indemnity insurance – and you can only get that if you don't make too many mistakes! So it is to constructors, who quickly point out if you have under-measured an item, and consultant architects/engineers, who likewise notice over-measurement not lost in tolerable "bunce", that I equally owe acknowledgement.

The best way to learn a subject is to teach it, and I readily acknowledge the many students I have taught, mainly at London South Bank University, but also at Reading University and Reading College. All my examples, including the examples in this book, have been developed, corrected and improved by students, often bringing into the class their knowledge and practice from outside. With 30 or so pairs of eyes scrutinising my work in lectures, any mistakes are quickly picked up and just as quickly corrected. To this acknowledgement must be added my fellow teachers at these institutions, who both provided and borrowed examples in developing our lectures.

Production of the book has been helped by Matthew Young, Peter McWilliam and Adam Dodgson of MTY Architects, who provided several of the photographs and developed the cover. They expressly disclaim liability for the architecture, which is my own, typical of a QS and not of the excellent architecture MTY produce! The orthographic drawings were produced using an academic version of the Bentley MicroStation CAD program, with their permission. Last, but not least, I must acknowledge my wife, Liz, who displayed infinite patience throughout the process.

Andy Atkinson
12 May 2023

Credits and references

Cover design

Peter McWilliam and Matthew T Young of MTY Architects, 25 Hunts Mill, Goldsmiths Lane, Wallingford, Oxfordshire, OX10 0DN, UK, based on author's drawing produced using MicroStation® CAD software and Sketchup Go visualisation program.

Photographs –

Figures 1.4, 1.13, 1.19, 1.20, 2.4, 3.11, 4.1, 5.3
MTY Architects

All other photographs

The author

CAD drawings and extracts

The author produced using MicroStation® CAD software. MicroStation is a registered trademark of Bentley Systems, Incorporated, 685 Stockton Drive Exton, PA 19341, USA

Extracts from the New Rules of Measurement, NRM2, Edition 2 –

Figures 0.10, 0.11, 1.5, 1.6, 1.25
Royal Institution of Chartered Surveyors, Parliament Square London SW1P 3AD, as accessed on 10th May 2021 at:

www.rics.org/profession-standards/rics-standards-and-guidance/sector-standards/construction-standards/nrm

Reference

RICS (2021) New Rules of Measurement NRM2, *Detailed measurement for building works*, 2nd Edition October, Royal Institution of Chartered Surveyors, London, UK, ISBN 978 1 78321 425 9

Tables

Figures

Glossary

abutment junction of two elements – for example, between a roof and a wall. The roof "abuts" the wall.
acrylic paint water-based paint consisting of pigment suspended in an acrylic emulsion.
adjustment, adjustments changing a quantity in the dimensions by making deductions or additions.
aggregate stones in gravel. Gravel used in concrete is graded into two stone sizes, (1) fine, containing small stones and sand, and (2) coarse, containing larger stones up to the maximum allowed aggregate size.
anding on (&) grouping items with identical quantities measured together. The two or more items are "anded" together.
annotation note added to a drawing to explain a feature.
architrave trim around a door opening to mask the joint between frame/lining and wall.
assembly a group of components with an overall unified purpose. For example, a balustrade assembly, an atrium assembly; often partially or fully prefabricated offsite.
back acting, back actor, back hoe excavator an excavation machine that uses a narrow bucket on an arm, working backwards to dig trenches. Small machines also have a broad shovel on the front end for reducing levels, excavating vegetable soil and moving bulk materials.
backfilling to trenches – earth or hardcore put back after foundations are constructed.
background material to which a fixing is made – categorised by NRM2 on page 45.
base material to which a fixing is made but outside the NRM2 classification for background on page 45.
bearing (for lintels) the part of a wall on which a lintel or beam "bears" or rests.
bedded set on a bed or backing of mortar or similar material.
bedded solid (tiles) fixed with mortar or adhesive without allowing any air pockets.
bending load load tending to bend a beam.
bills of quantities document containing in words all cost-significant items for constructing a project – includes measured work, preliminaries and provisional sums.
blank opening an opening left in a wall and the like for a later insertion – for example, a blank opening in a wall for a window.
blinding (1 – verb) associated with substructures. Treatment of a surface prior to covering with a sequential material. For example, blinding hardcore with sand to receive a polythene damp proof membrane.
blinding (2 – noun) a material used as a preparatory base for concrete or other materials. For example, rough concrete blinding for concrete ground beams, sand blinding for polythene.
bonnet hip tile a purpose-made tile resembling a traditional bonnet fixed over a hip rafter and bonded (interlocking) with adjacent plain tiling.

builder's work general work associated with a specialist installation. For example, builder's work in connection with heating installation.
built in (verb – building in) item or component placed in position, with adjacent work installed around it. For example, a window placed before masonry is built around it.
built-up roof roof construction built on site from basic timbers – rafters, ceiling joists and so on.
bunce colloquial term for slight over-measurement in taking off – usually intentional to avoid intricate detailed measurement.
butts hinges used for doors.
cavity closer a strip of material used to close a wall cavity at sill and jambs.
cement mortar mortar containing cement and sand.
centres the spacing of structural timbers measured from centre to centre of the timbers.
cold bridge a poor conduction path through an element, where the element generally is well insulated.
comb bedding placing a tile on a bed or backing of mortar or adhesive applied with a comb and, consequently, containing air spaces.
component a constituent part of a building which is fabricated as an independent unit. For example, a staircase, cupboard or glazed screen.
computer-aided design designs produced on a computer rather than with pen, pencil and paper.
contract sum the agreed payment for work at the outset of a construction contract.
contractor-designed works work intended to be designed and installed by a contractor, allowed for in bills of quantities by including a performance specification (employer's requirements) rather than detailed quantities.
coving plaster or plasterboard concave moulding at junction of wall and ceiling finish.
cross-section section or cut through an item in the shorter plane. Often related to orthographic drawings.
cure, curing allowing a material to strengthen over time. Associated with concrete, mortar and adhesives.
cut roof roof construction built from basic timbers – rafters, ceiling joists and so on – cut on site.
dabs gypsum plaster spots used to stick plasterboard dry lining to masonry walls.
damp-proof course (DPC) strip of impervious material laid on brickwork to stop damp penetration.
deduct (DDT), deduction reduction in quantity – the quantity is deducted from a previously measured quantity.
description wording of an item containing its unique specification. Only items with an identical specification will be included as a single item.
dimension check check of dimensions shown on drawings by designer, carried out by the taker off.
dimension paper specialised paper used for taking off quantities of building work.
dimensions (1) common term for the size of an object – in building work, often as represented on drawings.
dimensions (2) the written output of taking off as shown on dimension paper or in a computer file. Specifically relates to items of work and associated figures as opposed to waste calculations.
dims abbreviation of dimensions 2.
door lining the surround to an internal door opening, fixed after the opening has been formed.
door stop strip of timber set in a door lining to retain a door when closed.

dormers vertical windows let into a pitched roof such that the roof is extended over the sides and top of the projecting window.
double handling moving materials more than once in performing an operation. For example, moving earth to a spoil heap before removing from site.
double lap tile plain tiles which are lapped at least twice and three times at the ends of the tiles.
drawings visual representation of an object on paper at a reduced scale.
dressed, dressing turned up, down, under or into an element. Applies to sheet materials such as damp-proof membranes.
drip groove in underside of external timber components to cause water to drip off. Typically used under window sills.
dry lining plasterboard fixed to masonry walls as a finish, without the use of wet plaster other than as adhesive and to seal joints.
eaves edge of a roof adjacent to rainwater run-off.
eaves fascia board at eaves fixed vertically to ends of rafters.
eaves soffit board at eaves fixed horizontally to underside of rafters at ends.
eaves ventilator continuous grille and duct to provide ventilation to loft fixed over eaves fascia or to eaves soffit.
element several components or assemblies together performing a unified function in a building. For example, the external walls, roof and floors to a building are all separate elements.
elevations drawing showing vertical view of sides of an object. Associated with orthographic projection.
enumerated numbered.
estimate forecast of the cost of an item, component, element or building.
estimator person producing an estimate, usually for a construction contractor. Estimators may also carry out taking off for some types of building work.
facing (verb) the operation of presenting a brick wall that is visible.
facing bricks bricks intended to be seen. Of a better quality than common bricks.
figured dimensions dimensions (1) shown on a drawing.
final account final amount paid for a building contract after adjustments, for example, for variations.
finished (timber sizes) cross-section size of timber after it has been reduced (by planing) to the size specified.
first fix, first fixings services and joinery fixed before main finishings (such as plaster) are executed.
floating floor boarded floor laid loose over a material (usually sound or thermal insulation).
formation level level of ground below oversite at which construction starts. May be above or below the natural level of the ground and require that the ground be **made up** or **reduced**.
formwork moulds for concrete, often formed from ply lining on a supporting timber structure.
frame (door) surround to a (usually) external door. May be built in by masons/bricklayers or fixed to a blank opening.
frame tie strip of metal used for building a door or window frame into masonry.
furniture (ironmongery) door and window handles, knobs and so on.
gable (roof) roof with vertical gabled ends (often brought up in masonry).
gauged mortar mortar containing cement, lime and sand.
gauge for battens spacing of battens.
girth, girthing (1) perimeter of a closed element such as a wall to a building. May be measured on the extreme (outside) girth, centre line or minimal (inside) girth.
girth, girthing (2) total width of the surface of an open element such as a door frame, which may include width of frame, architrave, door stop and rebates.

grained, graining paint finish resembling timber grain.
grinning showing of underlying colour through paint finish.
groove semi-circular or similar cross-section routing to timber or similar material. A labour associated with timber frames.
ground bearing structural element gaining support directly from the ground. Applies to foundations and ground floor construction.
gypsum calcium sulphate dihydrate. The basis for widely used internal plaster. Forms the basic ingredient for prefabricated plasterboard.
half round ridge tile capping to apex of a roof of semi-elliptical section clay or concrete tiles about 450mm long, often bedded in mortar.
halved joint joint between ends of two timbers of the same cross-section consisting of a simple lap and rebate.
hardcore granular material (often broken bricks and stones) as a base for ground bearing elements.
head (of frame, lining or opening) horizontal top member.
herringbone strutting timber struts to floor joists set in a crossing pattern to resist transverse deflection.
hipped roof roof where the shorter sides are raked inwards towards the apex (usually) at the same pitch as the longer sides.
indent the corner on plan of a regular shaped building that has been indented inwards.
interlocking tile roof tile with edges that interlock to stop water penetration.
ironmongery operating machinery to a door, window or similar fitting. A door lock and furniture, for example.
item The part of a building being measured. This will be a distinct component of design and construction such as a concrete foundation, a brick wall or a timber window board. All items represent "**as built**" work such as they will appear in the finished construction, not the individual ingredients that go into making the item and follow the specialism or trade involved in constructing them.
item to measure note reminder left in taking off of an item yet to be measured (either by the taker off or another).
jambs (of frame, lining or opening) vertical sides.
joist hanger metal fitting to support floor joist at end.
key (1 – verb) prepare a surface for further application. For example, keying brickwork to take plaster.
key (2 – noun) surface prepared by keying.
labour a single treatment of timber throughout its length – for example, by rebating, grooving or weathering. Involves cutting a shape in the cross-section with a machine. A timber may have several labours worked on its profile.
lap (of sheet materials) extent to which sheet materials pass over each other at edges.
lap (of tiles) extent to which tiles pass over at ends. For plain tiles, lap is the extent to which tiles pass over to give three thicknesses of tile. For interlocking tiles, lap is the extent to which tiles pass over to give two thicknesses of tile.
latch door or window closing device.
laths strips of timber to support lime-based plaster.
left as laid work applied with no further treatment.
levelling, levelled making a component level in the horizontal or vertical plane.
lift, lifting tiles, plaster and so on separation of item from base. Often associated with causes such as dampness, frost or thermal expansion.
lime for building work calcium hydroxide; known as slaked lime.

lining visible surface treatment – door lining, dry lining and so on.
live load load imposed on an element other than the "dead" load of the element itself.
loadbearing partitions partitions taking a load from floors and/or roof.
made-up ground levels (noun) ground raised by filling with earth or hardcore to reach formation level.
make up ground levels (verb) raise ground by filling with earth or hardcore to reach formation level.
marking out indicating on building elements the position of services, such as plumbing or electrical installation.
measurable for taking off, an item required to be measured separately by NRM2.
measurement See taking off.
mensuration measurement of lengths, areas and volumes.
mesh steel reinforcement steel bars welded in two directions at right angles to give a mat/sheet and used to reinforce concrete.
New Rules of Measurement (NRM2) current agreement for the measurement (taking off) of building work in the United Kingdom.
nogging small timbers used between beams, joists and so on to give intermediate support to sheet material.
nominal (timber sizes) sizes before planing. Also referred as sawn timber sizes.
non-loadbearing partitions partitions dividing spaces but not supporting loads.
ogee profile based on classical orders of antiquity. Associated with timber sections and so on.
oil-based paint paint soluble in oil as opposed to water.
orthographic (projection) representing three-dimensional objects in two dimensions without perspective from fixed viewpoints, usually top and sides.
over-break making excavations bigger than they need to be, usually to avoid using earthwork support.
over-measure measure too much, rather than too little.
oversite treatment of the ground under a building to make a floor but excluding foundations to walls.
ovolo profile based on classical orders of antiquity. Associated with timber sections and so on.
party wall wall separating two dwellings.
pellet timber cylinder set into countersinking to hide a screw head in timber.
performance-based (specification) specification requiring a performance standard rather than prescribing a suitable solution.
pitch (angle) slope in degrees.
pitched sloping.
plain tile, plain tiling flat or cambered roof tile. In the United Kingdom, regulated to uniform size of 265 × 165 × 15mm.
plan horizontal view of an object based on orthographic projection.
plane (of work) extent to which work slopes, usually specified in NRM2 as horizontal, vertical or raking.
planing margin allowance on nominal timber for planing to give a wrought finish.
planking and strutting physical support members for earthwork support.
plan length (of timber) length of a timber as viewed on plan – foreshortened to the extent of the pitch of the timber.
plugging and screwing drilling a hole in a surface (usually masonry or concrete), inserting a plastic or similar plug and fixing with a screw.
pointing visible finish to joints between bricks.
preambles traditional name for a general specification written into bills of quantities.

preliminaries (or preliminary items) items of work related to a project that cannot be associated with measured work. Contained in "preliminaries" section of bills of quantities.
prime cost sum sum allowed for supply of goods, subject to adjustment in the final account based on the cost of the goods as supplied.
provisional quantity approximate quantity of an item subject to adjustment in the final account based on the quantities used in a project.
provisional sum allowance for work that has not been fully designed. Adjusted in the final account based on cost of work carried out.
quadrant small quarter circular section timber.
quantities the amount of the item. This is what is taken off the drawings.
query sheet request for information (RFI) from a designer in the form of written questions.
raking (adjective) at an angle. Associated with cutting bricks, tiles or other building materials at an angle – for example, next to the verge of a gable wall.
raking (verb) the act of scouring a material – for example, raking back joints in brickwork to form a **key**.
rate cost of an item per unit measure.
read understand drawings. To visualise two-dimensional information in three dimensions and relate it to finished building work.
rebate rectangular cross-section recess throughout the length of a timber.
reduce ground levels bring ground down to a formation level (as opposed to making up ground levels).
re-entrant part of a building on plan that enters in on itself, giving a semi-enclosed external space.
render and set two-coat plaster – thicker base coat of rough render and thinner fine coat of setting plaster.
reveals open edges of walls at doors or windows.
sash (or casement) opening part of a window. Two sashes slide over each other, whereas casements hinge open
sawn (timber) timber before planing smooth. Associated with nominal sizes.
scaffolding Access platforms and supporting structure. External scaffolding in the United Kingdom usually consists of a timber-boarded working platform, with galvanised steel tubes as support.
scale, scaled (1 – verb) reduce the size of an element by a fixed ratio (associated with orthographic projection).
scale (2 – verb) act of measuring with a scale ruler from a drawing.
scale ruler, or scale (3 – noun) ruler with divisions set to read to a reduced scale.
scaleable drawing that can be scaled as it represents elements reduced in size by a marked fixed ratio.
scaled taking off based on scaled, rather than written, dimensions on drawings.
schedules spreadsheets collectively itemising elements and their characteristics. Typical schedules include those for windows, doors, drainage and sanitary fittings.
screed cement and sand bed laid over a floor to take further finish.
second fix, second fixings services and joinery fixed after main finishings (such as plaster) are executed.
sections drawings showing vertical slices through a building. Associated with orthographic projection.
services installations required to operate a building. Includes incoming water, gas, electricity and communications as well as foul and surface water drainage.
set (door) assembly of door, lining and ironmongery, sold and delivered as an item.

set (lining or frame) assembly of lining or frame only, sold and delivered as an item.
side cast calculation made in preparation to entering dimensions in taking off.
signposts indication in dimensions of where in the project the measurement is taken.
single lap tile roof tile that only needs to be lapped at the end, as it has raised or interlocking edges.
site overheads See **preliminaries**.
SI units System Internationale units of measurement – most widely used international metric categorisation.
skin (brick) of hollow wall a brick wall built as part of a composite cavity wall. Each skin of a wall is measured separately.
skirting timber trim between floor and wall.
soffit underside.
span length of an opening beneath a beam or lintel.
specification detail in words of the nature of an item.
splashback tiles behind a sink, kitchen or bathroom unit.
splayed and rounded (architrave) two labours on timber, rounding one edge and splaying (sloping in cross-section) the timber.
spot bedding fixing components such as wall tiles on spots of mortar or adhesive.
spread of foundations extent to which a foundation is wider than the wall it supports.
squaring calculating quantities of item(s), whether in cubic, square or linear metres.
standard bill order order based on work sections of NRM2.
standard method of measurement agreement on what and how building work will be measured (taken off).
steel reinforcement bars of steel used to strengthen concrete.
stopped (labour) labour on timber that does not run through to the end of the piece but is stopped.
stopping material for filling imperfections in timber prior to painting.
stretcher bond staggered pattern of bricks or blocks based on the longest face showing externally.
summary total cost of a section of work – by page, work section, project and so on.
sunk weathering labour on timber involving cutting a sloping rebate. Associated with window and door sills.
taker off A person measuring building work. Also called a **measurement surveyor** or **quantity surveyor**.
taking off Also termed **measurement**. The process of measuring building work from drawings and also the product of this measurement. The product is also referred to as "**dimensions**" or a "take off".
taking-off list a list of items to be measured for a section.
taurus profile based on classical orders of antiquity. Associated with timber sections and so on.
templates (for windows/doors) rough frame of the same size as the window/door unit, built into masonry to ensure a good fit of the unit when installed.
tenon shorthand for a joint between two timbers at right angles, consisting of a mortice (hole) and tenon (projection) fitted together, glued and dowelled.
third round hip tile capping to apex of a hip rafter of semi-elliptical or less section of clay or concrete tiles about 450mm long, often bedded in mortar.
timesing multiplying in the context of taking off – waste calculations, side casts and dimensions. Represented by a/.
tongue, tonguing projecting edge to a timber, designed to fit into a corresponding groove in a second timber.

to take note reminder left in taking off of an item yet to be measured (either by the taker off or another).
trimmed joist joist cut back to form an opening.
trimmer joist joist supporting a trimmed joist where cut back.
trimming labour associated with forming an opening in a set of joists.
trimming joist joist supporting trimmer joists.
trowelled (bed) cement and sand bed laid over a floor to take a fine finish such as vinyl.
true to scale associated with drawings – one that represents scaled sizes accurately.
trussed rafters composite roof members forming a whole triangular assembly in framing a roof.
under-slating membrane membrane fixed under slates as secondary form of waterproofing.
under-tiling membrane membrane fixed under tiles as secondary form of waterproofing.
unit (of measurement) how an item is expressed dimensionally – m3, m2, m or No.
valley (in roofing) internal intersection of roof slopes lined with lead or purpose-made valley tiles.
verge edge of a roof adjacent to a gable.
waste calculation calculations in taking off leading to, but not forming, dimensions.
weathering sloping rebate in a timber cross-section.
wet plaster plaster applied unset on site (as opposed to dry lining formed from prefabricated plasterboard).
winder kite or triangular-shaped stair tread at intermediate change in direction of a staircase.
window board horizontal internal lining to window reveal at base.
work section division of work based on sections in NRM2 and loosely related to traditional trades.
work, working mortar applying material in use. For example, a bricklayer will work mortar.
working drawings drawings containing sufficient information for construction without the need for scaling.
wrought timber timber planed smooth.

Preface – how to use this book

Learning taking-off building work from drawings is very like learning many other skills. From bricklaying and plastering to driving a car, the process is much the same. Even learning to fly aircraft and complex surgical procedures adopt similar approaches. The skill is learned by a process of **demonstration**, **practice**, **knowledge of results** and **repetition**. A skilled practitioner, acting as tutor, demonstrates the correct approach; the learner attempts to replicate this; and the tutor gives feedback guiding the learner. The process is then repeated until proficiency is obtained and demonstrated.

Using this book follows the same incremental principle. Each chapter, with the exception of Chapter 8, is intended to be a self-contained demonstration of taking off a separate element. Read through a chapter, following the approach indicated, and then practice using the specification and drawings of a similar building provided in Appendices 5 and 6. The subject matter in each chapter changes, following the construction technology of each element of the building, but the approach remains the same. Once proficiency is obtained, the skill becomes progressively easier to transfer to novel elements and technology.

As it is easier to tackle taking off with the example and exercise resources laid out in one space, the drawings and specifications for both (Appendices 3–6) are provided on the Routledge website. This avoids having to dismantle the book to do the exercises. It is also envisaged that worked examples of the practice exercises, including some video presentations, will be provided to tutors and/or made available online in future.

This book is best tackled with a tutor providing knowledge of results and guiding further work. To tackle the exercises, you will need few special tools – a copy of the New Rules of Measurement (NRM2), a scale ruler and some traditional dimension paper. A set of colouring pencils and a phone or calculator would also be useful. It's best in learning taking off to use pen and paper rather than a specialist computer-based taking-off program, simply because it avoids needing to learn two skills at the same time, measurement and the computer program. Once basic taking-off skills are mastered, it is easy to apply this to a computer.

It is impossible to take off building work without understanding what is being measured, and taking off is usually taught at the same time as construction technology. A problem with learning the two subjects in parallel is that there is insufficient knowledge of technology prior to learning taking off. This book introduces a little of the technology before tackling each chapter, but this should be coupled with more detailed learning, perhaps by reading ahead of technology lectures. Also, although the book contains some photographs, as many site visits as possible should be made. Seeing actual construction in progress shows exactly what is being measured and accelerates acquisition of essential visualisation skills. It also makes clear the link between building work and arranging the means to pay for it. Overall, quantification of building work is an essential process in ensuring constructors are paid fairly for their effort and clients get value for money. I hope you enjoy using this book, learning taking off and putting the skill into practice in your future projects.

Andy Atkinson
12 May 2023

Introduction

The project

This book is a worked example of **taking off** building quantities for a traditional pair of semi-detached houses. This design of house is common in many countries but particularly popular in the United Kingdom. The design is still widely used, and, although small changes of detail have evolved, the basic pattern has remained the same for well over 100 years. The construction technology has also remained similar over this period, with incremental improvements in structural design, insulation and finish. Current construction has seen **components**, **assemblies** and **elements** increasingly fabricated off site. Prefabricated timber framed walling is now widely used (albeit often with a brick outer cladding), roof construction is of prefabricated trusses, ground floors are now often of precast concrete suspended beams and finishes are of **dry lined** plasterboard rather than **wet plaster**.

The construction for our houses is, however, essentially of site-based masonry and carpentry with tiled roofing, site-based internal finishings and services. In particular, the roof construction is site-based assembly of individual timbers, and the internal finishings assume traditional wet plaster walls. All these construction details are still current, in particular on smaller developments and for extensions and alterations (especially to older historic buildings). In addition, it is not necessary for construction technology to be innovative to learn taking off. The thought processes developed in measuring any construction, whether current or historic, are common to both, and it is easy to apply the process to any technology.

The documentation

Drawings

For taking off, a fairly detailed set of drawings is needed. They need not be full **working drawings** but must show all the cost-significant components of the building. Drawings are also needed for other purposes. In early stages, their primary use is to communicate with the client and with regulatory authorities. In the United Kingdom it is necessary to obtain planning permission, and the authorities ask for sufficient design detail to decide if permission should be granted. In addition, the authorities need to know if the design complies with technical building regulations, and this requires further details. As gaining regulatory approval will usually precede selecting a builder and agreeing on the contract, taking off will often be based on a set of planning and building regulations drawings. Occasionally, design will be further developed towards a full set of working drawings including a detailed specification, and this will give greater certainty to the taking off and resulting **bills of quantities**.

The drawings used for the example are fairly comprehensive but do not include all details necessary for precise construction. Further details will usually be illustrated on small-scale component drawings at scales of 1:20, 1:10 and 1:5. For most taking off, this level of detail is not

DOI: 12.01/9781003253129-1

necessary, but where relevant, either the taker off will ask for necessary drawings or, if not available, make provision for firming up measurement later (perhaps by including **provisional sums** in the bills of quantities to allow for the cost of undecided components). The set of taking-off drawings is shown in Appendix A3 and includes elevations, plans and sections, all at a scale of 1:50. The nature and use of architectural drawings is explained further in the following.

Specification

Drawings convey key information about what items look like (relating to spaces and positions) but have difficulty in communicating the nature of the items. For simple projects it is possible to provide some of this information in the form of **annotations** on the drawings, but this can quickly make them too cluttered. In addition, both the form and installation of modern materials and components require extensive detail that can only be conveyed in writing. As a result, for all but the smallest project, drawn information is supported by a separate written specification.

The specification, in turn, will refer to further technical detail. This technical detail is in national and international standards or in proprietary information provided by the manufacturers of materials, components and elements. Manufacturer's information will typically cross-reference national standards. As most of this specification information is standard, it is not necessary to refer to it in detail in taking off or in bills of quantities. The taker off only needs to provide cross-references to the relevant source as appropriate. Nevertheless, bills of quantities should contain enough information to allow an estimator to price an item initially without having to continually refer elsewhere. Many bills have specification details written into the document (often termed "**preambles**") or refer to a specification supplied at the same time as the bills of quantities.

For this book, the specification accompanying the drawings in Appendix A4 is extremely brief. This is to avoid over-complicating the example with what would be intricate standard detail. Just enough information is provided to allow the construction of workable items, satisfying in outline the requirements of **NRM2**. In the example, most of the items themselves assume there is a further detailed specification and cross-references are provided to this specification.

A beginner's guide to taking off

Taking off quantities is the **measurement** of building work from drawings in order to **estimate** a price. The **quantities** (volumes, areas, lengths or numbers of **items** of a building) are worked out, and, as a separate exercise, an **estimate** of the cost of the **item**, per unit measure, is calculated by an **estimator**. Finally, the **quantities** are multiplied by the cost per unit measure to give a total cost for the item, and all the total costs are summed to give an overall estimate, or price, for the work.

A simple example, using these terms, is foundation concrete. The taker off would measure the total volume of the concrete and present it to an estimator as shown in Table 0.1.

Concrete, being a fairly amorphous material (at least when unset!) has been measured by the taker off in cubic metres (m^3). The estimator will estimate the cost of making and placing the

Table 0.1 Example item in bills of quantities

Item	*Description*	*Quantity*	*Unit*	*Rate*	*Cost*
	Foundation concrete	50	m^3		

Table 0.2 Factor costs for foundation concrete

1	Materials	cement, sand, gravel and water
2	Labour	cost of a construction crew to make and place the concrete
3	Equipment	cost of a concrete mixer or other batching plant
4	Overhead	costs to cover general supervision and company expenses
5	Profit	return for investors in the construction company

Table 0.3 Priced example item in bills of quantities

Item	*Description*	*Quantity*	*Unit*	*Rate*	*Cost (£)*
1	Foundation concrete	50	m^3	100.00	5000.00

Table 0.4 Units of measurement

Item	*Description*	*Unit*
1	Foundation concrete	m^3
2	10-mm-thick floor tiling	m^2
3	150 × 25–mm timber skirting	m
4	1975 × 765 × 40–mm door	No

concrete per unit measure (i.e., per m^3) and put the answer in the **rate** column. To do this, the estimator will need to assemble costs as shown in Table 0.2.

Assuming the cost per m^3 amounts to £100.00, Table 0.3 shows Table 0.2 when priced.

The **unit of measurement** (how an item is measured – whether in m^3, m^2, linear metres or numbers) follows logic. Bulk items like concrete, excavation and filling are measured in m^3, and no sizes of the items are given in their description. Items covering areas such as roof coverings, floor tiling, wall plaster and brickwork are measured in m^2 giving the thickness of the items. Items involving lengths such as pipes, skirting boards, eaves boarding and kerb stones are measured in linear metres giving the width and height of the item (or, for round pipes, giving the diameter). Items such as doors and windows which are made, sold and fixed as discrete items are **enumerated** giving three dimensions (Table 0.4).

Which items to measure also follows logic. Most sections of building work (traditionally still often called "trades") have well-defined items of work following the nature of the material and the specialised method of construction for the trade. Most items also usually have a tangible presence – a concrete foundation, a brick wall, a tiled roof covering and a timber roof joist are examples.

However, sometimes less tangible items are involved, such as digging foundations, disposal of the earth arising from the excavations, compacting the bottom and holding up the sides of a trench or pit. Imagination and knowledge of construction technology are required to identify these items, and a decision has to be made whether to measure out each of these items separately or include one or more of the less significant items in larger, more important ones.

Whether an item is "**measurable**" is a practical question often troubling takers off. Do you:

- Include holding up the sides of excavations as part of the excavation item, both being carried out by the groundwork team, or measure two separate items?
- Measure out separately roof tiling, the battens holding on the tiles and the lining below the battens, all being carried out by roof tilers, or group them all together?
- Measure out separately a brick wall and pointing up the joints in the brickwork, both being carried out by bricklayers?

In many countries, both the *identification* and the *units of measurement* for items are fixed by agreement between takers off and estimators collectively, thus removing the need to decide. The agreement is presented as a **standard method of measurement** and is used as a convenient reference work guiding both the taker off in deciding what to measure and the estimator in knowing what to expect. The agreement is a guide, and identifying items and units of measurement primarily relies on knowledge of technology, logic and common sense, with the standard method being used as backup confirmation that they are being correctly applied.

Getting started

Reading orthographic drawings

As taking off is the measurement of building work from drawings, it is first necessary to know how to **read** (understand) drawings. Drawings are a visual representation of an intended building (or parts of a building) in two dimensions and to a size which can be handled conveniently. Although three-dimensional objects can be represented in two dimensions by using drawings showing perspective, it is difficult to do this and make the drawing **scaleable**, that is, to make it possible by increasing the size of drawn elements using a fixed ratio to represent the elements at full size so that the actual objects can be constructed.

To overcome this difficulty, three-dimensional objects are presented without perspective from fixed viewpoints. For buildings of regular construction, the viewpoints are from directly above (a **plan** view) and directly from every side (**elevations**). To give details of construction within the building, it is necessary to slice through and give "sectional" plans and elevations (usually termed simply "plans" and "**sections**"). Typically, plans are taken horizontally at about 1.20m above floor level (Figure 0.1), showing the position on plan of windows and doors, and vertical

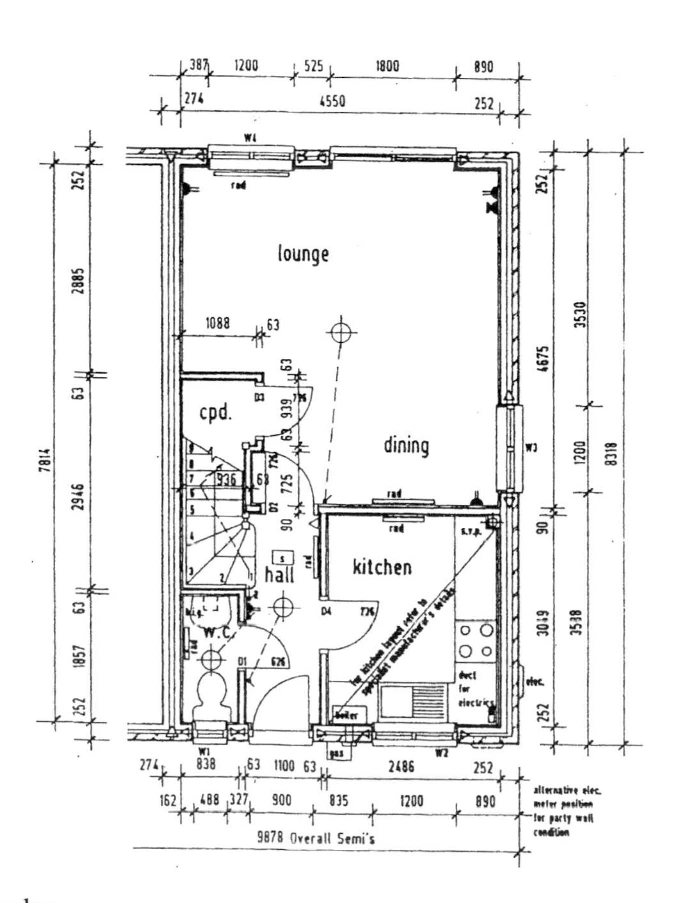

Figure 0.1 Typical floor plan

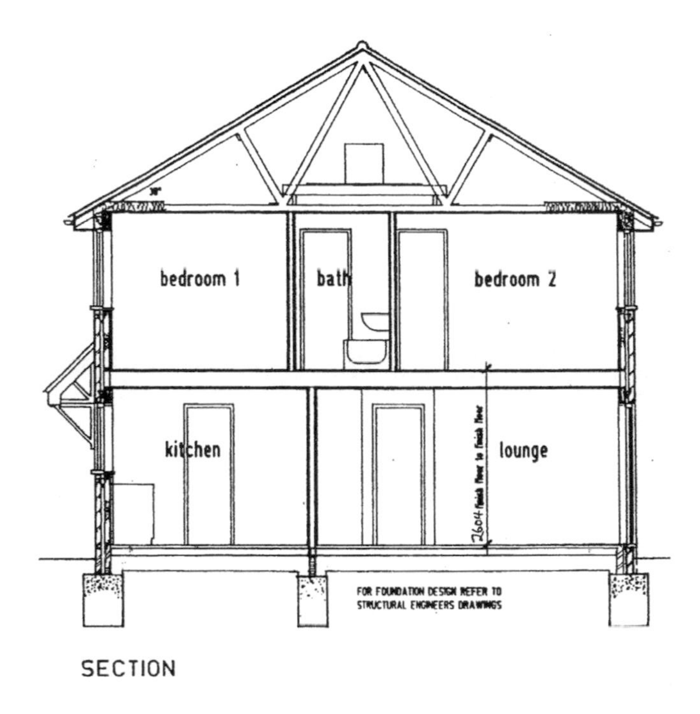

Figure 0.2 Typical section

sections are taken at positions which best show details of construction – for example, through a staircase or at the window positions (Figure 0.2). These types of drawings are known as orthographic projections and are used throughout architecture and engineering. Drawings are **scaled** to conveniently fit available paper – for example, A0, A1, A2 and A3 for use on a drawing board or on site and A4 for academic and office use. For **computer-aided design** (CAD), it is not necessary to scale drawings whilst on the computer (the programs can work in any size), but the **scale** will be selected on printing paper drawings. Scaling is linear – for example, at 1:50, 1 millimetre represents 50 millimetres in length. Typical scales are 1:200, 1:100 and 1:50 for layout drawings showing a whole building and 1:20, 1:10 and 1:5 for drawings showing smaller component or assembly details. Other scales are used for more specialised circumstances such as for overall planning of buildings at 1:1250 and 1:2500.

When drawings are reproduced strictly to scale, it is possible to measure off them and use the **scaled** measurements for taking off and on site. To do this, it is necessary to use a **scale ruler** (known simply as a **scale**), which has pre-marked divisions of actual sizes reduced to the scale size. **Scaling** from drawings in this way is not always accurate, as the drawing itself may not be running **true to scale** (printing to PDF files often reduces image sizes to fit paper, so prints are not true to scale), and architects will write dimensions on the drawing to make this unnecessary. Where drawings have been "fully dimensioned" and show all dimensions necessary to fix the positions of detailed components to the extent that they can be built without scaling, they are known as **working drawings**. Taking off is not usually from working drawings but an earlier version (in order get on with the job and to save time), and one preliminary task in measurement is to check that a drawing is running true to scale. This is done by scaling any marked dimension on the drawing (or a "scale indicator", if there is one) and ensuring the scaled answer is the same as the **figured dimension**.

Handling drawings

The production of building information, including drawings, has a dynamic character that is often not apparent in descriptions of architecture and construction. Drawings for all but the smallest project are developed in stages, with sketches preceding outlines, which precede drawings for estimating and contract formation, which in turn precede working drawings. Information is changed and amplified as drawings approach construction.

Taking off usually takes place somewhere between the production of outline and contract drawings and is based on information that is subject to change before the contract is placed. Accurately identifying by date and reference the drawings used for taking off is, therefore, vitally important. A preliminary, if rather basic, task in taking off is to check through, record and date-stamp all drawings to be measured. If drawings change subsequently and time allows, the changes can be incorporated in the taking off. If there is no time or opportunity to incorporate changes, at least the date of taking off can be identified, and it will be clear if changes have been made subsequently. Receiving drawings electronically removes the need for a physical date stamp, but clearly it is still necessary to accurately date files and folders without overwriting them with subsequent information.

Dealing with drawing errors

Whilst perusing drawings it is also common to identify missing information or errors on the drawings and in any accompanying specification. The taker off performs a useful service in picking up these errors and reporting them back to the architect and should formalise the process of checking by producing a **query sheet** (Figure 0.3). A query sheet will identify the error or omission and the date of query. It will also identify any answer given, who gave it and when. In practice, experience allows the taker off to save time by making assumptions about many details (for

Andrew R Atkinson PhD MSc FRICS

Job Number

Job Title

Section

Taker Off

Sheet Number

Ref	Query	Date	Answer	Date	Designer

Figure 0.3 A query sheet

example, based on information from previous projects) and carry on with taking off whilst waiting for answers. Where these assumptions are correct, the architect will follow them and amend the drawings to fit the taking off.

Completed query sheets provide summary written evidence of decisions made during design development. Feeding completed sheets back to the architect prior to placing a contract ensures that changes to drawings are implemented and design information from drawings and taking off is the same.

A specific check that can be carried out on drawings is an arithmetical check of the architect's figured dimensions, that is, that all runs of dimensions in the same direction through a drawing add up to the same figure. Working plans are often "through dimensioned" – lines of dimensions are given, for example, for the overall length of the building from left to right, the lengths of all walling and windows over the same overall length and the lengths of all internal spaces and partitions over the same overall length. Each run of dimensions should add up to the same figure, and the taker off can check this. CAD has made it more difficult for architects to make errors in dimensioning, but it is still possible for dimension runs to be started or stopped at the wrong position, and errors can result. Identifying these errors not only assists the architect but also ensures that measurement is not based on erroneous dimensions. Figure 0.4 shows a fully dimensioned simple plan for the substructure to a building, and Figure 0.5 shows part of the associated dimension check for the same building.

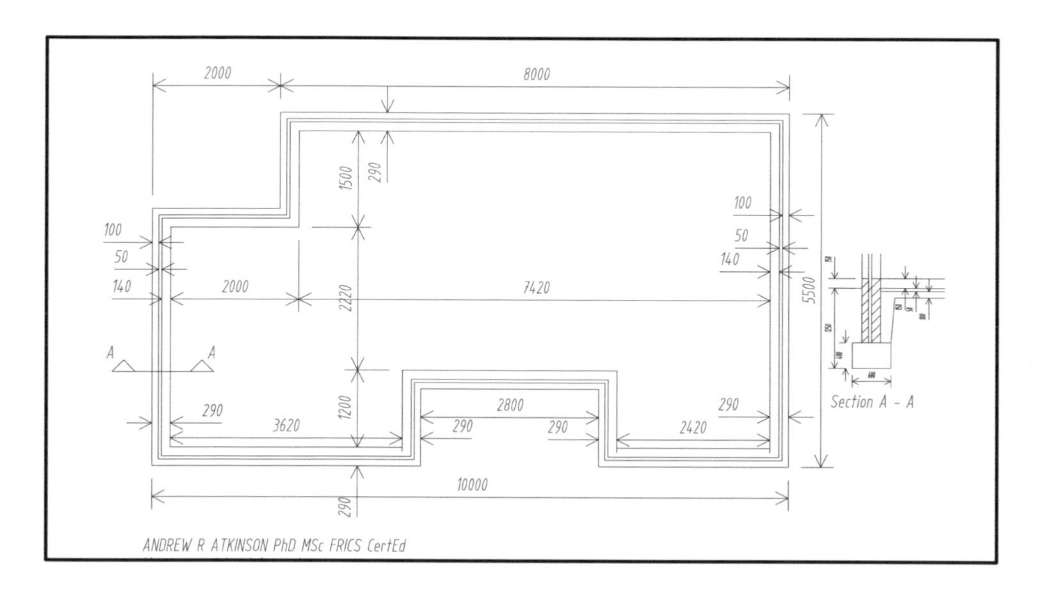

Figure 0.4 Simple dimensioned floor plan

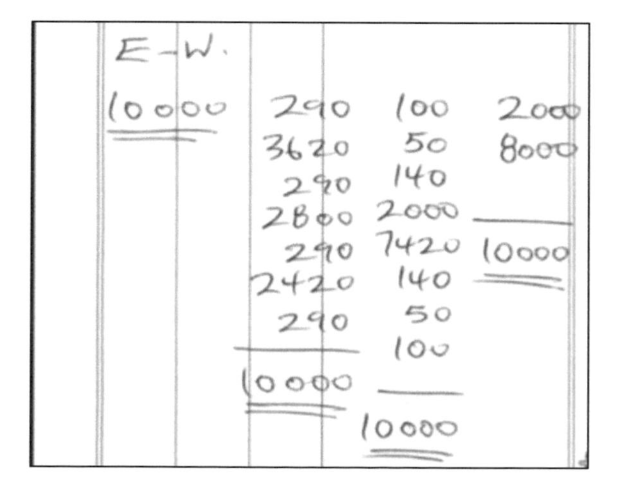

Figure 0.5 Example dimension check in handwriting

Identifying items to measure

Fully and completely identifying every item to measure is one of the harder tasks in measurement, and the complete omission of items is the biggest cause of taking-off errors. In comparison, errors in **mensuration** (getting measured lengths wrong) are relatively insignificant.

Identifying measurable items is helped by colouring drawings using a suitable convention. There are not now any standard colours for different materials, but commonly used colours are red for brickwork, blue for blockwork, green for concrete and yellow for timber. The colours adopted for each material are not important, but the act of colouring is. It requires attention to construction detail and allows rationalisation of measurement prior to starting taking off. Identification of items is also helped by constructing a list of items to be measured. Although experienced takers off may not do this, during training, lists can be useful and, when checked with examples of past similar work and any standard method of measurement, will help to identify items.

Colouring drawings, carrying out dimension checks, date stamping drawings and constructing query sheets all force a pause in starting work, allowing the overall configuration of the building to be appraised and helping with the identification of items.

The conventions of taking off

On the face of things, it doesn't matter how you take off building work as long as the answer is correct and arrived at speedily! The parameters of the right answer quickly are keys to any skill and distinguish the skilled practitioner from the amateur. Conventions, although the origins may be ancient and obscure, are designed to help this. Conventions also establish a common language between similar practitioners, allowing ready understanding and efficient transfer of information. Accordingly, even when the reason for a convention is no longer apparent, it continues to be used to aid communication.

To illustrate this justification, the product of taking off (known as **the taking off**, **dimensions** or just **dims**) in many instances is not only used by the person producing it. For larger projects, many takers off will be working in parallel and may need to share information or take over work started by another. Dimensions are also regularly used as reference documents when changes are made during construction. A post-contract surveyor or administrator may need to refer to original dimensions in order to value changes and must be able to read the calculations and descriptions, including locating the position in a building of the original taking off. It does not matter what the convention is, but it must be consistent, in much the same way as a language is consistent, to allow efficient communication.

Form

Output documentation – bills of quantities

Quantities need to be translated to a form that can be easily understood and priced by estimators. Estimators need to find items readily and build up prices logically from basic to more complex. As the actual building work is carried out by specialists, the first categorisation within an output document is, therefore, by specialist work section (trade), then by discrete specification within the section. Items are then categorised by whether they are in cubic, square or linear metres and number in that order. For similar items within this categorisation, further division is in ascending order of cost. Pricing is easier if identical items only appear once in the pricing document, although computer production makes multiple pricing much

Table 0.5 Layout of bills of quantities paper

Item	*Description*	*Qty*	*Ut*	*Rate*	*£p*
1	Excavation starting 150mm below existing ground level; foundation excavation not exceeding 2.0m deep	29	m^3	10.00	290.00
2	Disposal: excavated material off-site	27	m^3	12.00	324.00

easier. In many countries, the output document following these conventions is **bills of quantities** and is produced in a form that can be priced by estimators from several builders quoting for a project. The process for internal documentation within a building firm is similar, but the form of the bills of quantities will be less rigidly defined, perhaps by not strictly following any standard method of measurement. The form of bills of quantities is very similar to that illustrated in Table 0.3 and gives an item number, description, quantity and unit of measurement, with two further columns for a rate and item price to be inserted by the estimator. Appendix A2 shows the measured work section of the bills of quantities for our project, and Table 0.5 is a small extract from the bills. This shows items for excavating foundations with a quantity of $29m^3$ and disposal of excavated material off site with a quantity of $27m^3$. Note: there is slightly more excavation than disposal in this extract, as some excavated material is **backfilled** (put back) around the foundations.

Input documentation – dimension paper

Taking off will not usually be carried out in the same order as presented in bills of quantities, as it is easier and more logical in taking off to follow an elemental sequence rather than a trade order. An elemental sequence divides a building into architectural elements rather than a specialist trade division within a construction sequence. Elements include substructure, external walls, internal walls, roof construction and covering, windows and doors and so on. Many trades may be involved in an element, and this gives rise to the need to re-order the taking off, once completed, into the order of bills of quantities. With paper-based systems of taking off, the re-ordering process is quite intricate, involving manually collecting up and summing all identical items and transposing the results onto bills of quantities paper. Computer quantities have efficiently automated this process. Coding items and entering the code and dimensions into the computer allows reformulation of the result to traditional bills of quantities or many other useful formats. The program will also carry out required calculations, such as **squaring** the dimensions and any preliminary calculations required by the taker off.

The paper-based look of dimension paper has survived, by and large, the computerisation of bills of quantities, and most computer programs display input "pages" as if they were pages of dimensions. The common language characteristics of taking off, mentioned previously, explain this. The printed or screen-based format of modern dimensions is familiar and based on time-honoured precedent, but the justification for this format other than as a common language is unknown. It was in use, but not explained, at least as early as 1888,[1] in much the same way as presented in Figure 0.6. Dimension paper is divided vertically in two, a right and left side, and within one side into four columns. The paper is used vertically, with the taker off working down one side and then down the other, and the columns are as shown in Figure 0.6. Figure 0.6 is an extract from the taking off for our project and shows, in part, how the quantities in the bills were obtained. The squared quantity (shown as **29.46** in the squaring column) for foundation excavation is the same as the quantity in the bills shown in Table 0.5. Table 0.6 explains in more detail the use of the four columns in traditional dimension paper.

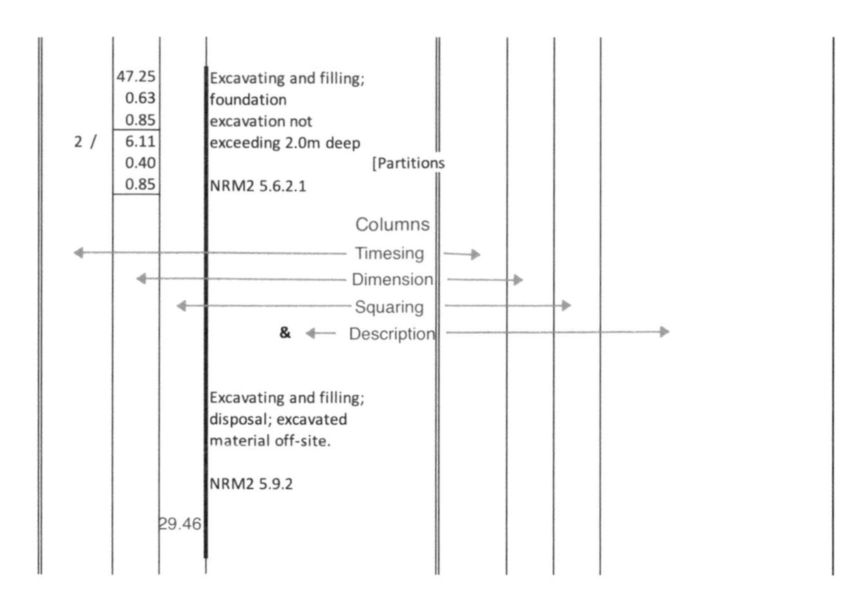

Figure 0.6 Use of columns in dimension paper

Table 0.6 Use of columns of taking off paper

Column	*Use*
Description	To describe the item in sufficient detail for pricing. A traditional test of the sufficiency of a **description** is whether the item could be sketched from it. Key elements of a description include outline of specification and sizes of the item, where not in the dimensions and parameters required by any standard method of measurement. As much of the specification is standard to many items, it is excluded from the item description and placed in a standard specification or in headings preceding the item. Modern methods of measurement (such as **NRM2**) and computer programs provide a defined structure and language for items, removing the need for freehand "composition".
Dimension	To carry out the taking off, involving measuring lengths, breadths and widths entered vertically in the dimension column to two decimal places with a line under each completed dimension. Cubic items involve three dimensions entered vertically, followed by a line. Superficial items involve two dimensions entered vertically (the third dimension being in the description), and linear items involve a single dimension entered vertically (the second and third dimensions being in the description). Numbered items are written in the same way as linear items, but, being discrete items, without decimal places and with all three dimensions given in the description. Dimensions are "stacked" to allow the taker off to add or deduct volumes, areas and so on from several locations. So a single set of dimensions for an item may combine multiplication, addition and deduction (subtraction).
Timesing	To multiply up a dimension by a given factor. Repetition of dimensions is common in taking off, and the **timesing** column allows a dimension to be "timesed up" by however many identical items there are of the dimension measured. The timesing figure is followed by a diagonal stroke (similar to internet "slash" notation and, rather confusingly, the usual division stroke). Timesing can also be timesed – for example, the first timesing might be for the number of dimensions to the floor of a building, the second timesing for the number of floors to the building and the third timesing for the number of buildings in a development. More on the reasons for and elaborations of this intricate method of showing timesing calculations is given in the following.
Squaring	To calculate the net quantity of the item. Traditionally this is always called "squaring" even if the item is cubed, linear or enumerated. The taker off will not normally actually do squaring, it being considered an unskilled task. Even where the taker off is charged with the task, he or she will always square up after all the measurement is finished in order not to disturb the train of thought necessary for taking off. Computer quantities provide instant results – as dimensions are entered, squared results are automatically generated.

Entering dimensions

The dynamic nature of construction, mentioned previously, is relevant when writing dimensions. The finished taking off is a snapshot of the quantities of the building or element at the date of measurement, but the dimensions are likely to be used by subsequent participants. In the *first* instance, discrepancies between measured and actual quantities may be noted by others and it is necessary to track down, in the dimensions, where and what quantities were taken. In the *second* instance, other parties (including both architects and constructors) regularly rely on output from bills of quantities to identify what is actually to be built where. In the *third* instance, variations are required to the work and it is necessary to return to the dimensions to take out work to be omitted or change the details of other work.

The taker off is, therefore, required to leave a *clear audit trail* that can be followed by others. Taking off is like writing a novel rather than a textbook, and the "thread" of the "taking off story" should be clear. This requirement drives the method of entering dimensions and is another reason for the form of taking off. A summary of the general rules for taking off is given in Table 0.7, and their application is demonstrated throughout this book.

Table 0.7 General rules for taking off

Direction of flow	Down one side of traditional dimension paper and then down the other.
Spacing of items	Liberal spacing of items to allow forgotten items to be inserted in the correct expected position. A general guide is four items per side of A4 paper. This is less important on a spreadsheet or computer program, where items can be cut and pasted around.
Work from dimensions to descriptions	That is to say, write dimensions out first and then construct the descriptions. This order will allow rationalisation of measurement to give the least number of repeated items.
Stack dimensions to each item measured	Stacked by product and sum.
Use **To Take** items and lists	That is, a note of items that cannot be measured yet – perhaps information is missing, or it is part of a different element. This will ensure that items are not completely forgotten.
Write in ink	To force maintenance of an audit trail. Less possible with spreadsheets and computer programs – all input is effectively in pencil.
Write all calculations	Do not carry out calculations mentally – these cannot be checked by others or the computer.
& on (**Anding on**)	For sets of items with identical dimensions & on the subsequent item below the first
Deducts	For deducted dimensions, write these after all the positive "added" items or write a separate "Deduct" item.
Line through items in a taking-off list as they are measured	To maintain record of what has been taken off. If using spreadsheets or a computer program, highlighting items measured is an alternative.
Order of dimensions	Always length × width × height as shown on the drawing
Signposts	Add locational and similar information to dimensions to aid those subsequently using the taking off.

A simple example in action – taking off plastering and painting

Taken from the main example, Figures 0.7 and 0.8 show two floor plans of our project as used throughout the book. Figure 0.9 shows a small extract of taking off based on the two floor plans. The format is on traditional dimension paper, but the extract was actually produced on an Excel spreadsheet. The subject matter is the plastering and painting of internal walls (marked up in red on the plans), carried out by the skilled trades of internal plasterer and decorator, respectively. Although these trades are carried out consecutively, they form the same constructional element of internal finishings and are easier to measure together.

It is not necessary to fully understand the technology of internal finishings to follow the example. Most people are familiar with painting internal walls to dwellings and have come across plaster work of some kind, but an outlinc of technology follows.

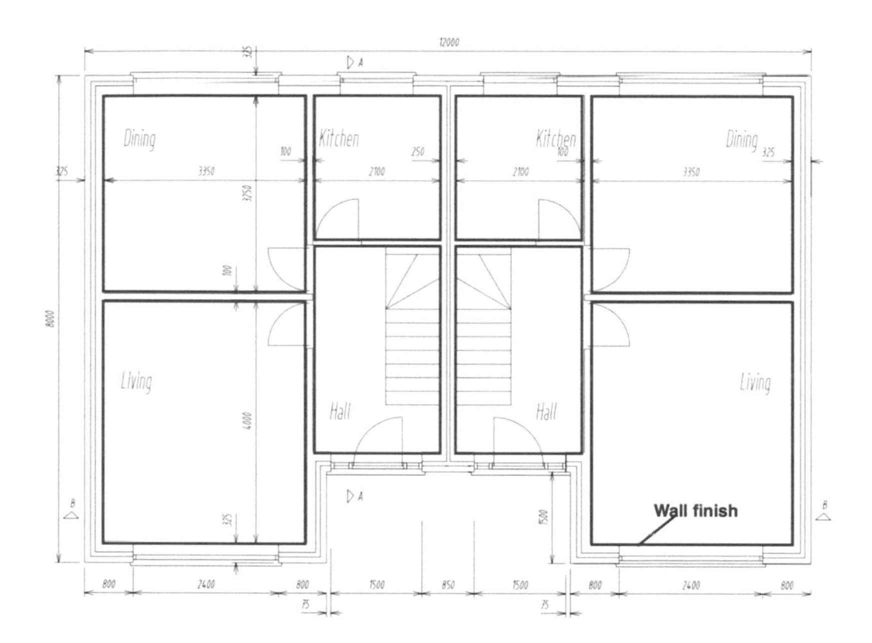

Figure 0.7 Ground floor finishings plan

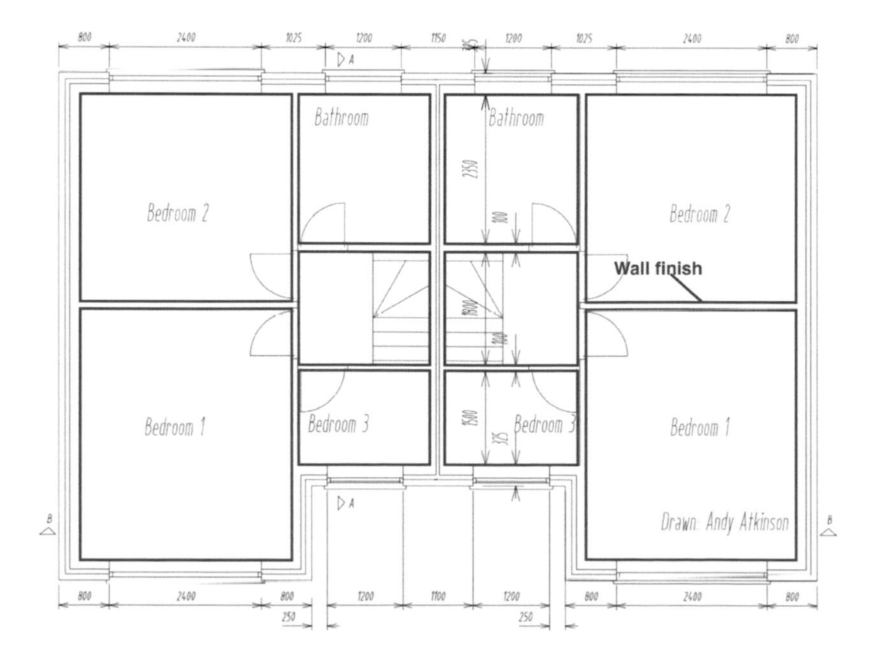

Figure 0.8 First floor finishings plan

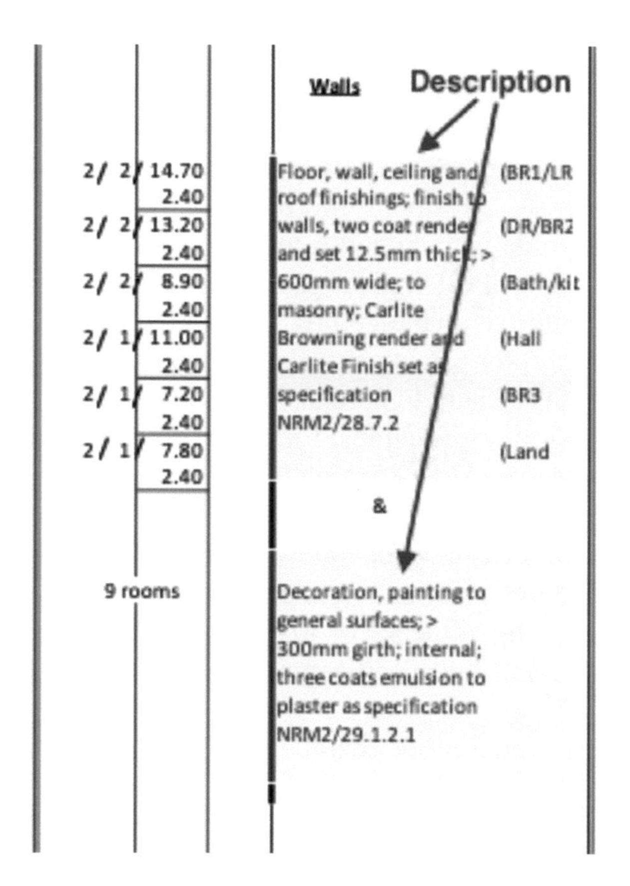

Figure 0.9 Extract of finishings taking off

I have first considered the plastering. Most ordinary internal plaster is applied to masonry walls in two coats, consisting of a base (or "render" coat) of thicker work (about 10mm) designed to take up the unevenness of the underlying structure and a finish (or "setting" coat) about 3mm thick giving a flat fine surface. In modern work in the United Kingdom, the finish coat is **gypsum** based. The render coat may also be gypsum based mixed with sand or, for a harder impervious plaster, cement based. Historically, **lime** (calcium hydroxide) finish and base coats were often used, sometimes with a third intermediate "floating coat" between the "render" and "set". In restoration and historic repairs, lime plasters are still widely used.

Painting to plastered walls usually involves two or three coats of emulsion paint – a paint consisting of a polymer suspended in water, which dries to give a durable coating. Some specifications for applying emulsion require a watered preliminary coat to counter the rapid absorption of the paint into the dry plaster, and a third coat may be specified for new work to avoid the base colour of the plaster "**grinning**" (showing) through. For rooms with high condensation, a waterproof emulsion, an acrylic or oil-based paint, may be specified. Historically, whitewash (lime suspended in water) may have been used, sometimes made washable by the addition of oils. Again, whitewash may be specified in restoration and repairs.

The descriptions of items in the example broadly follow the technology. The first one starts by explaining what section of work is being measured – *Floor, wall, ceiling and roof finishings*. Clearly the section covers all types of finishings, but only finishings to walls is relevant to the part being measured. So, the next phrase narrows things down – *finish to walls*. The next phrase deals with the specification in broad outline – *two coat render and set 12.5mm thick*. This description is familiar from the technology outlined previously. The following phrase, "> *600mm wide*", is simply stating that all the areas measured with this item are at least 600mm wide. The reason for this categorisation is that it is more difficult (and consequently more expensive) to apply plaster in narrow widths of less than 600mm. If narrow widths of plaster are encountered, then they will be measured separately (as to who says they must be measured separately, see the role of rules of measurement subsequently). The next phrase states the base to which the plaster is applied – *masonry*. The nature of the base determines the difficulty of applying the plaster and, again, the price. Thereafter, the phrases are giving more details of the specification (Carlite Browning and Carlite Finish being trade names) and refer the reader on to further detail in a separate specification. The specification will often give extensive instructions on application and refer to information provided by a materials manufacturer. In summary, therefore, the item should give an estimator enough information for pricing and also, incidentally, buyers and construction managers considerable information relevant to purchasing and application.

The painting item follows similar principles. First the broad category of "*Decoration*" (of all types), then "*painting to general surfaces*" (which would include surfaces such as walls, ceilings, doors and panels), then "> *300mm girth*". As with plastering, narrow surfaces are more expensive to paint than broad, so they are measured separately. "*Internal*" refers to the location of the work in relation to the building, and this information is given as painting is more expensive *externally* than *internally* both because of the influence of weather and difficulty of access. The base of the painting is given as "*plaster*", and the specification is three coats applied in accordance with a separate specification.

The categorisation and description of the two items follows the technology and logic of carrying out the work – the general price-sensitive conditions under which it is carried out – and both the broad and detailed specification for the items. However, although largely following logic and common sense, the categorisation is not left for the taker off to determine. It is pre-set in a standard method of measurement – in this example, the RICS **New Rules of Measurement NRM2 edition 2** (RICS 2021). The references given at the end of the items are to the relevant categories in NRM2, so, for the first item, if the reference 28.7.2 is looked up in NRM2, the appropriate description can be constructed. Figures 0.10 and 0.11 show the relevant extracts from NRM2. The coding given in the item from NRM2 can therefore be traced back to the rules – NRM2, Section 28, item 7, level one descriptor 2.

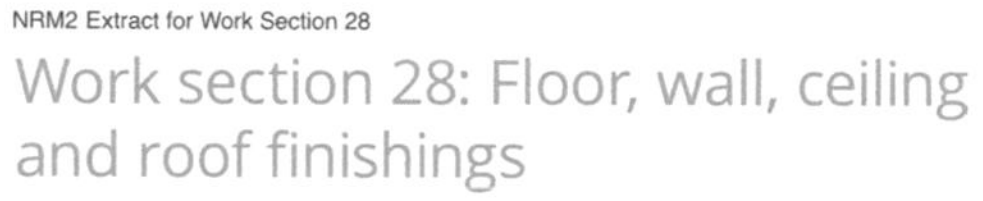

NRM2 Extract for Work Section 28

Work section 28: Floor, wall, ceiling and roof finishings

Figure 0.10 NRM2 extract for Work Section 28

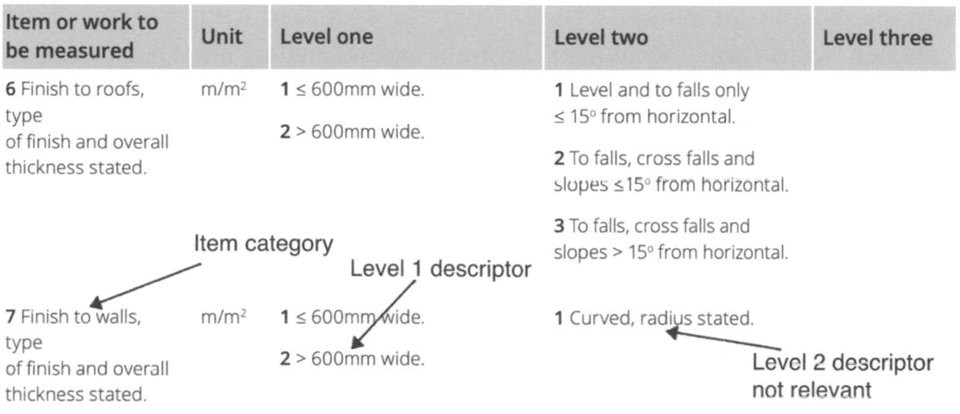

Item or work to be measured	Unit	Level one	Level two	Level three
6 Finish to roofs, type of finish and overall thickness stated.	m/m²	**1** ≤ 600mm wide. **2** > 600mm wide.	**1** Level and to falls only ≤ 15° from horizontal. **2** To falls, cross falls and slopes ≤15° from horizontal. **3** To falls, cross falls and slopes > 15° from horizontal.	
7 Finish to walls, type of finish and overall thickness stated.	m/m²	**1** ≤ 600mm wide. **2** > 600mm wide.	**1** Curved, radius stated.	

Item category

Level 1 descriptor

Level 2 descriptor not relevant

Figure 0.11 NRM2 extract for Work Section 28, finish to walls

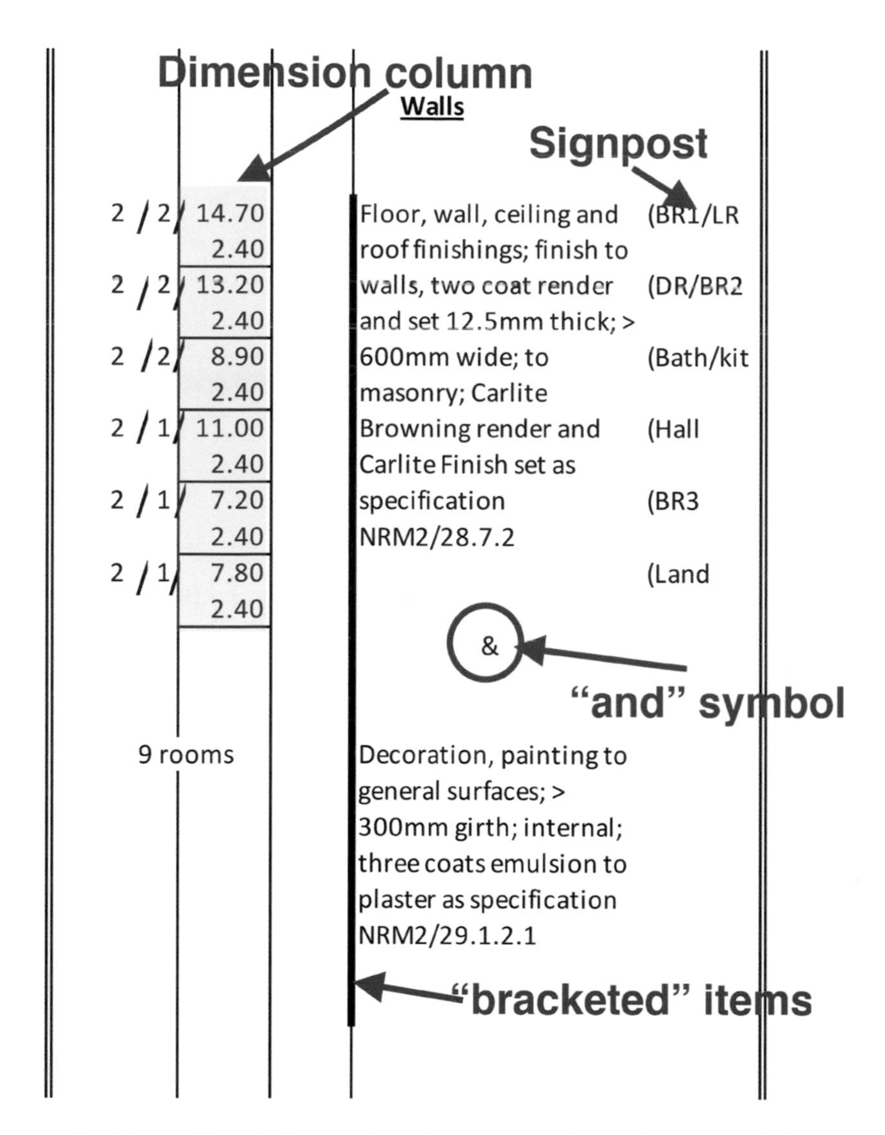

Figure 0.12 Extract of taking off with dimension signposts and bracketing highlighted

The dimensions are given room by room, and the taker off has previously worked out the perimeter of each room, requiring each perimeter to be multiplied by the room height (in this example, all heights are 2.40m). To the right of the description are abbreviated signposts – for example, "(BR1/LR". Context and experience would tell a taker off that this means "bedroom one" and "living room" – very useful information if work is omitted to only these rooms or the specification changes during construction. Figure 0.12 shows the taking off with the dimensions and signposts highlighted. The full list of abbreviated signposts in this instance is given in Table 0.8, and the dimensions can be traced back to the rooms shown on the two floor plans.

The areas of plaster and decoration are identical, so the taker off has "**anded**" on the two items using the ampersand symbol "**&**". **&** means that the dimension applies *both* to the plaster and the decoration. To make this absolutely clear, the taker off has also "bracketed" the two items (the thick vertical line immediately left of the descriptions).

The timesing column has been used to multiply up identical rooms. The first set of timesing (labelled "Timesing level 1" and highlighted in blue closest to the dimensions) reflects the signposts – for example, dimension one is for "BR/LR1", and the timesing is "2". Dimension four is for "Hall", and the timesing is "1". Arithmetically, it is not necessary to multiply a dimension by 1! However, the taker off has deliberately done so to indicate that there is only one room of this dimension in this set of timesing. The second set of timesing (labelled "Timesing level 2" and highlighted

Table 0.8 List of signposts for wall plaster

Signpost	*Full description*
BR1/LR	bedroom 1 and living room
DR/BR2	dining room and bedroom 2
Bath/Kit	bathroom and kitchen
Hall	hall
BR3	bedroom 3
Land	landing

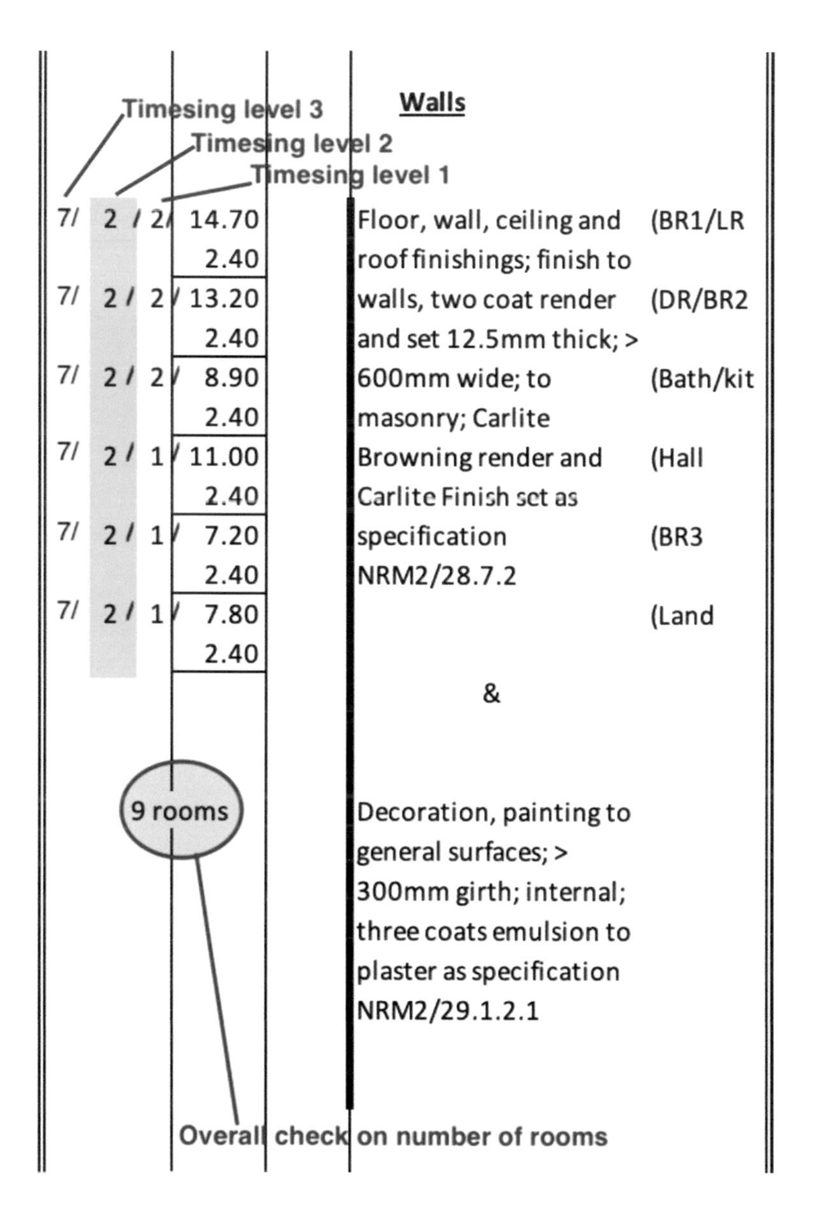

Figure 0.13 Extract of taking off showing use of timesing column

in red) further from the dimensions is all figure "2". Arithmetically, the taker off could have simply used the figure "4" for the first dimension but, again, has deliberately deconstructed the arithmetic to two timesing figures, 2 and 2. In this instance the drawings indicate that there are two identical houses in the building being measured (semi-detached dwellings), and the taker off wants to make it quite clear that there are two rooms per dwelling and two dwellings per building. If there were seven buildings on a development, then the taker off would further times up the figures by providing a timesing of "7" to the left of the current set, as illustrated in yellow in Figure 0.13.

At the bottom of the run of dimensions is the figure "9", which is the sum of the number of rooms in one dwelling. In taking off it is easy to leave out elements of construction – rooms being decorated, floors being concreted, drainage chambers being excavated and so on. Leaving out elements, whilst not as fundamental as leaving out whole items, is a major cause of error. The taker off has an eye not only on the dimensions being measured but also on the overall number of elements involved. This is also another reason the dimensions are further timesed up by the number of dwellings in the building and why, should there be seven buildings in a development, there would be a further timesing up by 7. The "9" is an overall check that all rooms have been included. Showing the timesing up in a layered, deconstructed way allows the taker off to check for accuracy and anyone to follow the dimensions later on in the project. The way the descriptions, dimensions and timesing have been entered clearly shows that the taker off is not only getting the correct answer but is showing how the figures have been arrived at – writing a novel, not a textbook!

The dimensions so far suggest that there are no windows or doors in the rooms being measured, but clearly these must exist, as they are shown on the plans. The taker off is measuring overall, with the expectation that an **adjustment** will be made deducting areas not plastered or painted. The process of measuring overall and deducting for voids is illustrated in Figure 0.14, where the plaster is deducted for openings. The item is the same as the main "addition" item in Figure 0.9 but written in abbreviated format, with the suffix "as before" added to indicate that it has been measured earlier. The item is also headed DDT to indicate that it is a deduct (and a reduction in

Timesing	Dimensions	Squaring	Description	
			Adjustments	
			Wall finishes	
2/ 1/	2.40		DDT	(LR
	1.50		Render and set to	
2/ 2/	1.20		masonry walls as before	(Kit/bath
	1.05			
2/ 1/	2.40			(DR
	2.10			
2/ 1/	1.50			(Hall
	2.10			
2/ 1/	1.20			(BR3
	1.20			
2/ 2/	2.40			(BR1/2
	1.20			

Figure 0.14 Extract of taking off showing deductions for windows and doors

quantity) and not an additional quantity. Manually, or by computer, the net quantity of plaster will be adjusted downwards in the final bills of quantities.

For most taking off, not only is measuring overall and deducting easier, it is also safer. It is better to forget to make a deduction and thereby over-measure than to fail to make an addition. So, in measuring the area of a simple building, the taker off would start by measuring the whole area and adjust (deduct) for any **indents** and **re-entrants** in the building.

This brief extract shows pretty much all the common conventions of taking off. Although the conventions are difficult to comprehend initially, they do not vary for any of the elements measured in this book. Provided the taking-off example is carefully followed and applied through further exercises, they will quickly become familiar and proficiency will be achieved in their use.

Key points covered in this chapter:

- The nature and purpose of taking off
- The information basis for taking off – drawings and spccification
- Reading orthographic drawings
- Preparing for taking off
- The conventions of taking off and bills of quantities production
- Form of bills of quantities and taking-off (dimension) paper
- The role and use of standard methods of measurement and **NRM2**
- A short example of taking off

Note

1 Banister Fletcher (1888) *Quantities*: A textbook for surveyors in tabulated form, Batsford, London United Kingdom.

1 Substructure

Take a look at the foundation and ground floor plans, sections A-A and B-B. These show the nature of the substructure to our two houses. The construction is traditional and involves strip concrete foundations which obtain their support directly from the ground. On these are constructed block cavity outside and party walls and loadbearing partitions. These extend upwards to the superstructure masonry. The ground floor is of concrete, also obtaining its support directly from the ground, but built over a layer of granular material (called **hardcore**), which is laid to ensure an even level base for the concrete.

To measure the work involved in constructing the substructure, the taker off would follow more or less the construction sequence, so a preliminary task would be to identify the sequence of work and assess the items that need to be measured.

Some items are not specific to the substructure and cannot really be measured with it. For example, managing the work, although relevant, is not specific to the substructure and needs to be spread over the whole project. Similarly, setting up the building site, providing site accommodation and facilities affect all elements of construction, not just the substructure. Even setting out the foundations cannot really be associated with just the substructure, as the engineers are also involved in more general managerial tasks and several sites. These items, often called "**site overheads**" or "**preliminary items**", have a cost but would be measured and included elsewhere than in the substructure measurement.

Some items are excluded from the substructure for simplicity. In particular, it will usually be necessary to prepare the site before starting by demolishing buildings, digging up old paving, cutting down trees, removing vegetation and hedging. In this example, it is also assumed that the site is quite flat and the ground does not need **levelling** to give a surface upon which to build. The work involved in installing **services** such as drainage, electricity, gas and communications is also excluded in order to concentrate on the main measured items.

After allowing for items that cannot be measured, as they are too general to be in the substructure, and for items left out for the sake of simplicity, a list of items identified by the taker off would be as shown in Table 1.1.

Preliminary activities

AA1 to AA12 of the example is the full take-off, produced on a spreadsheet, for the substructure. In modern practice, all but the simplest measurement would involve the use of specialised taking-off software rather than a general spreadsheet program. However, for learning purposes it makes little difference if examples are constructed using a taking off program, a spreadsheet or, indeed, using "old-school" pen and paper as illustrated in Figure 1.1.

Computer quantities save time and effort in writing descriptions, sorting output to bills of quantities order and carrying out arithmetical calculations, but the thought process of taking off

DOI: 12.01/9781003253129-2

Table 1.1 Taking-off list for substructures

Item	*Work involved*
Excavate vegetable soil	Stripping off vegetable soil, as it is not of bearing capacity and may have value if sold. Vegetable soil is earth with a high proportion of organic matter, such as roots and decayed plants.
Disposal of vegetable soil	Taking the vegetable soil off site or to a specified location on site.
Excavating trenches	Usually by employing a **back-actor** bucket on a mechanical excavator guided by a laser.
Disposal of excavated material	The same volume of excavated material will be disposed as has been excavated. For general soil, this is usually off site, but a small proportion will often be put back over the foundations when complete. This will be dealt with later in the list and in the taking off.
Earthwork support	Holding up the sides of the trenches. The nature of support is chosen by the contractor but could be traditional **planking and strutting**, sheet piling or no physical support.
Concrete foundation	This type of bulk concrete, not reinforced with **steel reinforcement**, can be poured straight into the trenches against the earth.
Block skins of hollow walls, block partitions and brick facings	This is the masonry in the foundation walls, some of which is in **skins of hollow walls** (i.e., forming part of a cavity wall) and some of which is in partition foundations. Each skin is measured separately. A small amount of external masonry will be in **facing bricks**, as this will be visible on the face of the building.
Forming the cavity between skins of hollow walling	There is not much cost associated with forming a space between walls except providing ties to hold the skins together.
Concrete cavity filling	More expensive than foundation concrete, as it needs a finer **aggregate** (stone) to fit in the cavity and is more difficult to form.
Damp-proof course	A strip of impervious material provided between substructure and superstructure masonry to stop damp rising.
Earth backfilling to foundations	The trenches outside the building can often be filled in using the earth dug out in the excavations. As this will be the same material as excavated, the volume will *not* be removed from site but backfilled instead.
Hardcore backfilling	Filling an imported stable granular material back into the inner side of trenches. It is not possible to return earth dug out from the foundations, as this will not be sufficiently stable and compacted to provide a base for the ground floor.
Hardcore bed under the concrete floor	This will use a similar material to that for the hardcore backfilling but will need to be **blinded** with a finer material such as sand or fine gravel to give a smooth surface as a suitable base for following work – in this case a polythene damp-proof membrane.
Concrete bed	This bed does not contain steel reinforcement and can be poured directly onto the polythene membrane.
Tamping concrete	Involves providing a mechanically formed compacted level surface on the concrete.
Damp-proof membrane	An impervious material, such as polythene, to stop damp rising through the floor.

remains the same. This process involves visualising in three dimensions construction features shown on drawings, capturing these features as measurable items in text form and quantifying them to enable pricing by an estimator. Well-constructed items can be "reverse visualised" back to the three-dimensional object by the estimator, taker off or other interested parties.

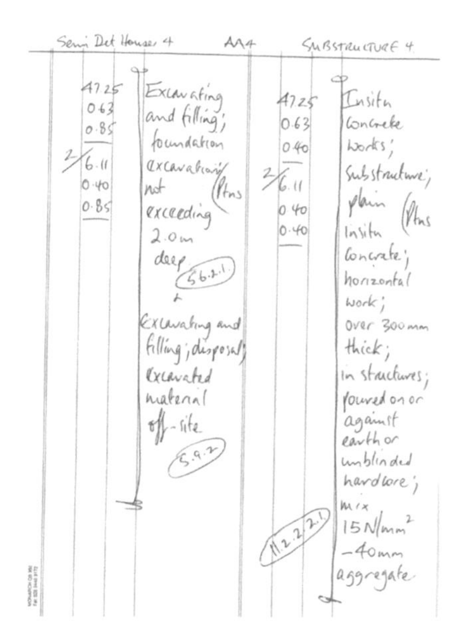

Figure 1.1 Handwritten taking off

Whether paper, spreadsheet or program based, the taker off will need to identify him/herself on the media and assign a job number and serial number for the element being measured. The example, although produced by spreadsheet, assumes a paper-based output, so pages and sections are numbered in series. The identity of the taker off is important in case there are questions in the future, so the initials of the taker off are added as a prefix to the page number. The drawings used, with dates and revision references, are identified in the taking off so that subsequent revisions can be tracked. Drawings are coloured up as an aid to their perusal and to identify items. To deal with preliminary questions to the designer arising from the perusal, the taker off will prepare a query sheet, with questions formally entered and dated. A detailed **taking-off list** is useful in training and would be included formally on an early dimension sheet so that it can be worked through during the measurement. If taking off is from dimensioned drawings, the dimensions will be checked, where possible, and the calculations written onto the dimension paper.

So, in summary, prior to starting measurement in earnest:

1 Date stamp or otherwise record all drawings received.
2 Select relevant drawings from the set.
3 Colour up drawings to identify items and aid perusal
4 Construct a query sheet for initial and subsequent queries.
5 Sign the first sheet of dimensions and initial each subsequent sheet.
6 Identify job number and section on each sheet.
7 Construct a taking-off list and include this on the dimensions (or in notes to a data file).
8 Where possible, carry out dimension checks to identify any errors and aid perusal of drawings.

Figure 1.2 shows the layout of page 1 of the substructure taking off following the previous guide. Although conventions will vary between different practices and locations and between methods of production, the approach and essential identifying information are common to all.

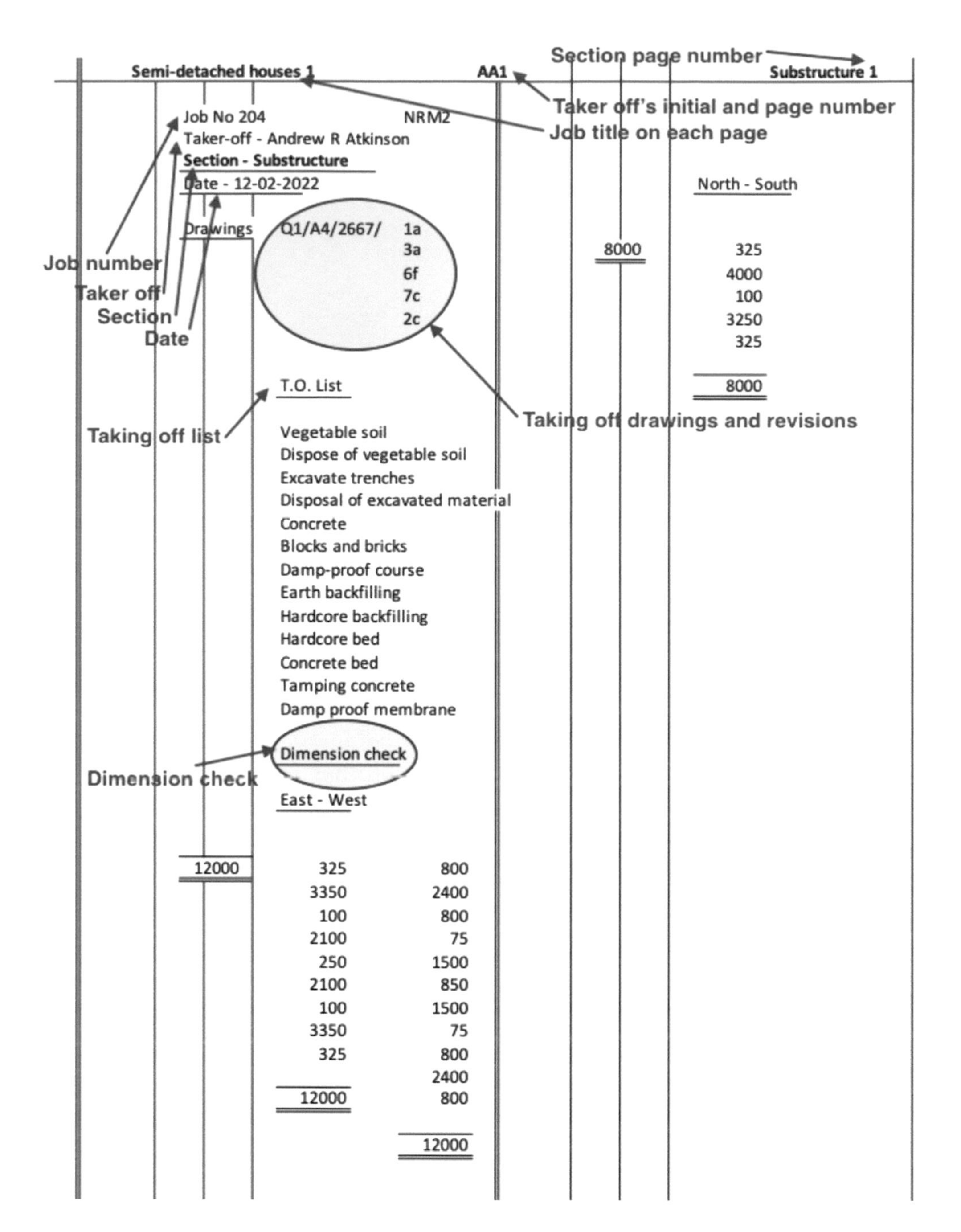

Figure 1.2 Page 1 substructure taking off

The **dimension check** is based on the orientation of the building as east–west from left to right and north–south from top to bottom – the designer has placed a compass north point on the Ground Floor Plan. In the absence of a compass point, any suitable identifying method can be used, such as orienting features on the site.

To follow the dimension check, Figure 1.3 shows the ground floor plan annotated with the lines of dimensions picked out in blue (horizontal) and red (perpendicular). So, the east–west first overall dimension of 12000 is the top blue line of dimensions. The middle blue line starts with 325, 3350, 100 and so on through the building, and the bottom blue line starts with 800, 2400, 800. The north–south dimensions adding to 8000 are entered similarly, and, as would be hoped, dimensions sum up to the same figure for each direction.

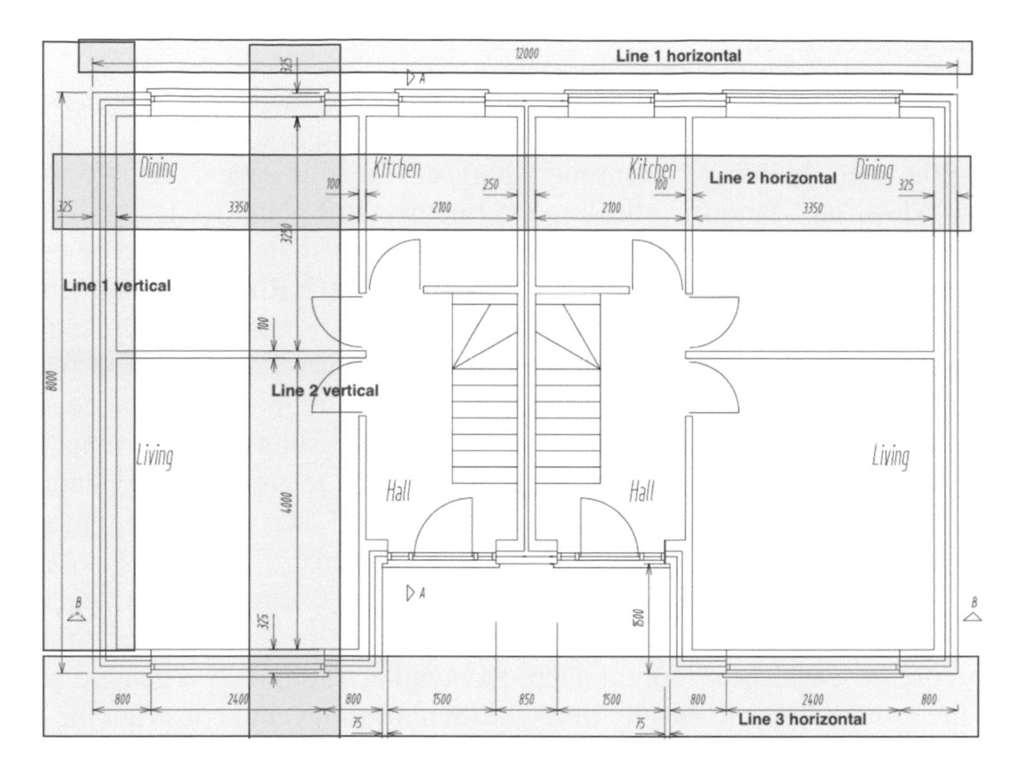

Figure 1.3 Lines of dimensions on ground floor plan

Figure 1.4 Topsoil excavation

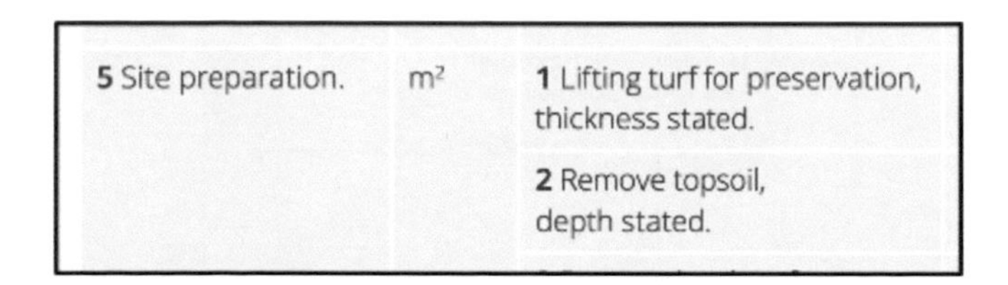

5 Site preparation.	m^2	**1** Lifting turf for preservation, thickness stated.
		2 Remove topsoil, depth stated.

Figure 1.5 NRM2 topsoil excavation

Excavating the vegetable soil

Consulting the taking-off list, the first item is excavating vegetable soil. The specification for "Topsoil" is "150mm vegetable soil to be excavated, retained and spread and levelled on site 150mm thick, 20m from building". From the taking-off list, this appears as two items of "vegetable soil" and "disposal", broadly reflecting the two items of plant needed – an excavator and a truck or dumper. It is necessary to determine exactly how these items should be measured. It is here that reference is made to the **New Rules of Measurement (NRM2)** – the agreement between takers off and estimators on how items will be measured and presented. The relevant section of NRM2 for excavation is 5.5.2, as shown in the extract at Figure 1.5. In the Excavation section, removing topsoil is classed as Site Preparation and measured in m^2 stating the depth of excavation. The relevant section of NRM2 for disposal is 5.11.1.1.1, as shown in the extract at Figure 1.6, and for shallow filling of less than 500mm deep is also to be measured in m^2 stating the depth. So it appears that 150mm of vegetable soil is to be stripped off, taken 20m from the building and spread on the site to exactly the same depth as it was excavated.

An extract from Section A-A is shown in Figure 1.7, and the excavations for vegetable soil and trenches are coloured pink. It can be seen from the section that the ground level is the same as the top of the hardcore bed, which is 150mm deep. Excavating 150mm of vegetable soil will bring the surface of the ground down to exactly the right **formation level** to construct the hardcore bed and concrete floor. This coincidence is very unlikely in practice, as there would normally be some need to **make up** or **reduce ground levels** to get to the correct depth for construction.

Having determined the items to be measured, that they are to be presented in square metres and the depth is 150mm, it is now necessary to determine the area of excavation. The drawings and dimension check have shown the building to be 12000 × 8000 overall (dimensions on drawings in the United Kingdom are shown in SI units of millimetres), so an initial measurement of these two dimensions would be a good start. It is better to **over-measure** initially and reduce (or **deduct**) for unwanted **indents** or **re-entrants** afterwards. That way, if an adjustment of quantity is forgotten, it is more likely to involve too much of an item – a less costly and embarrassing error than having too little.

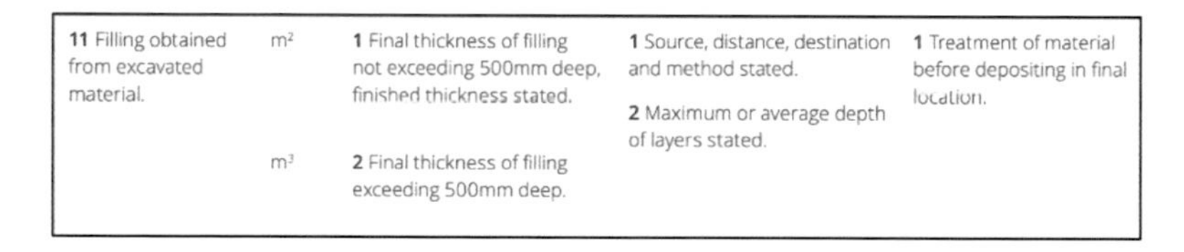

11 Filling obtained from excavated material.	m^2	**1** Final thickness of filling not exceeding 500mm deep, finished thickness stated.	**1** Source, distance, destination and method stated. **2** Maximum or average depth of layers stated.	**1** Treatment of material before depositing in final location.
	m^3	**2** Final thickness of filling exceeding 500mm deep.		

Figure 1.6 NRM2 filling excavated material

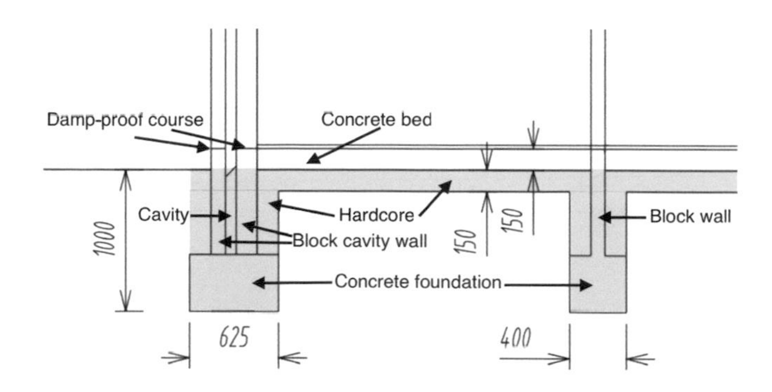

Figure 1.7 Topsoil and trench excavation section

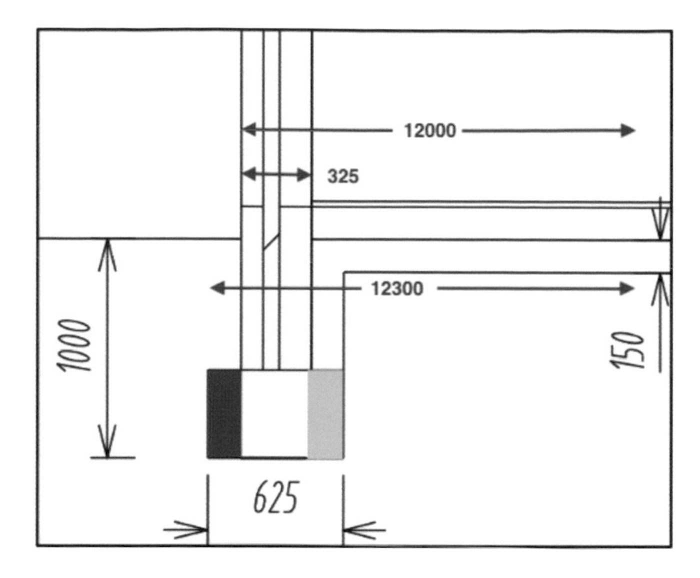

Figure 1.8 Spread of foundations

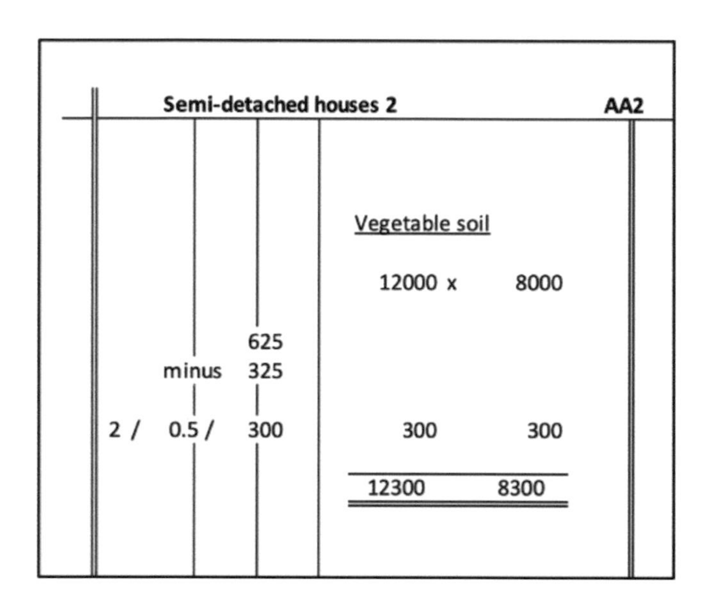

Figure 1.9 Topsoil waste calculation

Dimensions of 12000 × 8000 are, in fact, too little for the amount of vegetable soil that needs to be excavated, because they do not allow for the **spread** of the foundations beyond the wall. Look at Figure 1.8. The dimension to the outside of the wall is 12000, but the spread of foundations, shown in red, means that more needs to be excavated. The width of the spread can be calculated by taking the width of the foundation and deducting the thickness of the wall (i.e., 625 – 325 = 300). This gives the spread on both sides of the wall – red on the outside and blue on the inside – so the answer of 300 needs to be divided by 2. Finally, this extra is required on both sides of the building, meaning that to get the overall extra, the resulting figure must be multiplied back up by 2. So, the overall length × breadth of the vegetable soil excavation is 12300 × 8300.

Page AA2 of the dimensions (Figure 1.9) shows this calculation as it would appear using taking-off conventions. The calculation is known as a **side cast** or **waste calculation** to distinguish it from the dimensions themselves. Three features of the calculations are:

1 All calculations are entered expressly so that the figures and method can be checked.
2 Waste calculations are in mm.

3 The working is shown – each step of the process is noted – for example, 300 is multiplied by 0.50 (to show it is one half of the overall spread of foundations) and then multiplied by 2 (to show the answer applies to both sides of the building).

This method of systematically setting out calculations with workings showing each step allows for checking of the arithmetic and for any future taker off using the dimensions to follow the process of taking off. This is particularly valuable if it is necessary to make changes and re-measure work during the construction phase. It avoids needing to completely repeat taking off where small changes are made.

The overall area of 12300 × 8300 is too much, as there is a re-entrant area at the front of the building. The size of this area is 4000 × 1500, as shown on AA2, the horizontal width being calculated by summing the dimensions across the area. However, this width is *reduced* as a result of the spread of the foundations, as shown in Figure 1.10. The net figure becomes 3700. Note the perpendicular N–S depth of the area is unchanged at 1500mm, as the extra spread of foundations at one end is compensated by an identical spread at the other end. A mathematical way of presenting this fact is that the internal and external corners of the re-entrant area at the spread of foundations, outside face of the wall and inside face of the wall form a parallelogram on which opposite sides are equal. So the dimension remains at 1500mm in all cases.

Having determined the overall dimensions for the vegetable soil and the deduction for the re-entrant area, it is now possible to enter these as items and dimensions, as shown on AA2 (Figures 1.11 and 1.12).

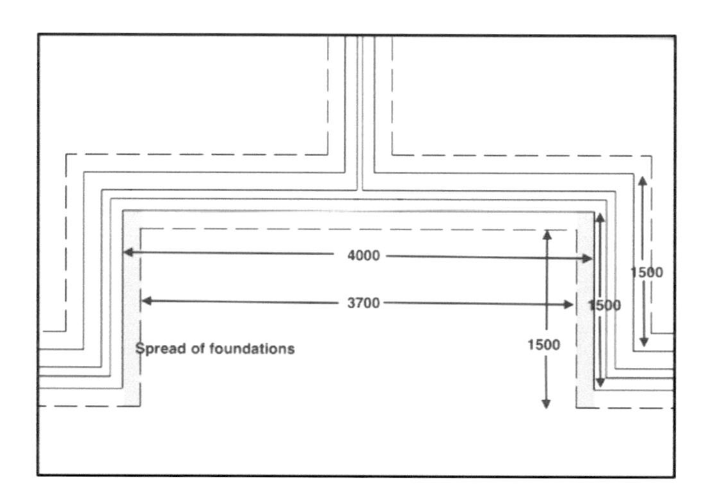

Figure 1.10 Re-entrant at front of building

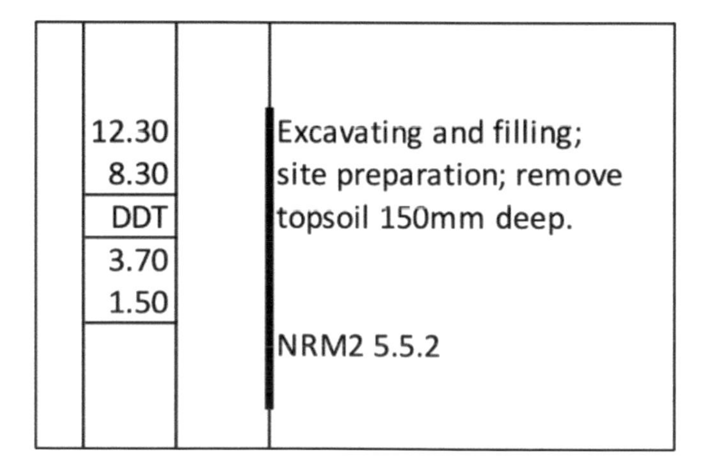

Figure 1.11 Topsoil item in dimensions

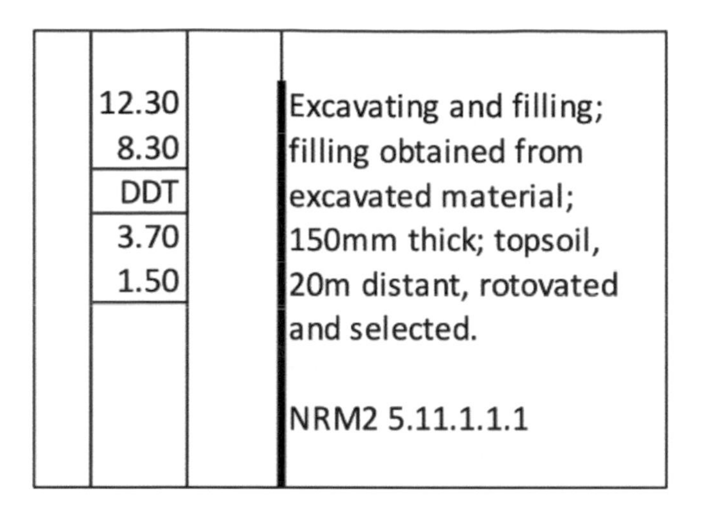

	Dimensions		Description
	12.30		Excavating and filling;
	8.30		filling obtained from
	DDT		excavated material;
	3.70		150mm thick; topsoil,
	1.50		20m distant, rotovated
			and selected.
			NRM2 5.11.1.1.1

Figure 1.12 Topsoil disposal item in dimensions

Figure 1.13 Foundation excavation

The overall dimensions are entered as square metres (two figures vertically, then a horizontal line), with the deduction for the re-entrant area immediately below. The item is constructed directly from NRM2, and the NRM reference is provided below the description as a guide to the learner.

The disposal item is dealt with similarly, and, in this case, the area for disposal is the same as the area for excavation.

Note: to get soil from an excavation to its final resting place 20m distant, it may be necessary to move it more than once. For example, the soil may be initially placed in a temporary heap whilst work is in progress and only moved, spread and levelled when all other work is completed. Using an intermediate stage for disposing of excavated material is known as **double handling**. NRM2 does not require that double handling be measured, except when it is expressly required by the architect. Any double handling will still need to be priced into the item, and this will be done by the estimator but *not* measured by the taker off. Double handling may be expressly required for vegetable soil when the architect has yet to decide on the final position of the earth and may instruct that it be heaped temporarily until a decision is made. If that is the case, then the disposal to a spoil heap and subsequent spreading on site will both be measured.

Excavating the foundations

The next items on the taking-off list are *excavating trenches* for the foundations and *disposal of excavated material*. Excavating trenches is a more precise activity, using more complex techniques than excavating vegetable soil, and is consequently likely to be considerably more expensive. Subsoil is specified as being removed from site. As the vegetable soil has been removed previously, all the trench excavation will be in subsoil and removed from site. The two items are therefore different from the vegetable soil items. This is confirmed by NRM2 (5.6.2.1 for the excavation and 5.9.2 for the disposal).

To calculate the volume of trench excavation, it is necessary to work out the trench lengths, breadths and depths. Breadth and depth are easily determined from the drawings. All depths are the same, and there are two breadths, 625mm for the outside and party wall trenches and 400mm for the internal loadbearing partition trenches. The correct length of the outside wall trench is along its middle, as marked by the white dotted line in Figure 1.15. This can be calculated first by working out the length on the line of the outside of the trench. This is the same line as the edge of the vegetable soil excavation, and we know that the area of vegetable soil is 12300 × 8300mm overall. If these two figures are added and the result multiplied by two, it gives the length all around the building (i.e., this is the same as taking two sides at 12300 and two sides at 8300 and adding the four sides together). The re-entrant part of the building at the entrances is 1500mm deep, so a further two lengths of 1500mm need to be added to give the overall length. The dimensions involved are marked on Figure 1.14. The overall length works out at 44200, but this is too long – the correct length is along the white line in the middle of the trench shown in Figure 1.15. To get this figure, for each corner, the length needs to be reduced by a distance amounting to half the width of the trench on each side. Figure 1.16 shows the white line being reduced in this way – once in the north–south direction and once in the east–west direction. This occurs for every corner, and, as most buildings are of a regular shape, there are four corners in total.

We can disregard the corners at the re-entrant, as the reduction at the outer corners is compensated by extra for the inner corners (Figure 1.17). For nearly all simple rectilinear buildings,

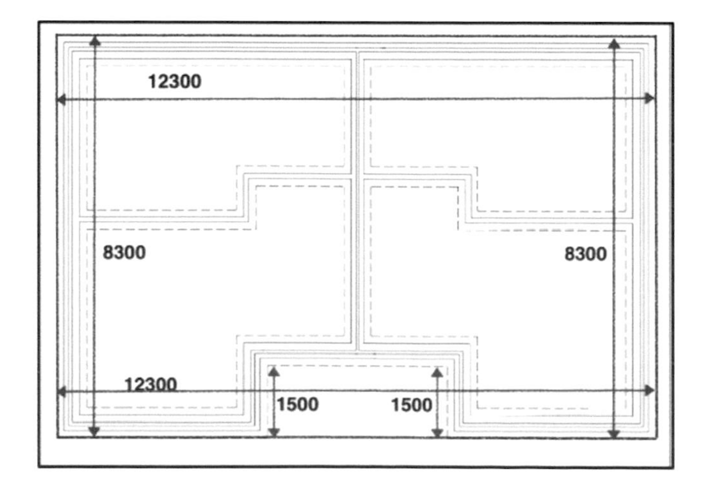

Figure 1.14 External dimensions of topsoil area

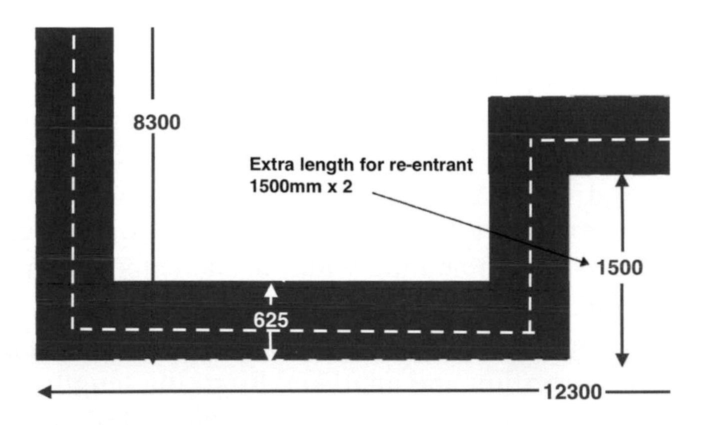

Figure 1.15 Centre line of trench excavation

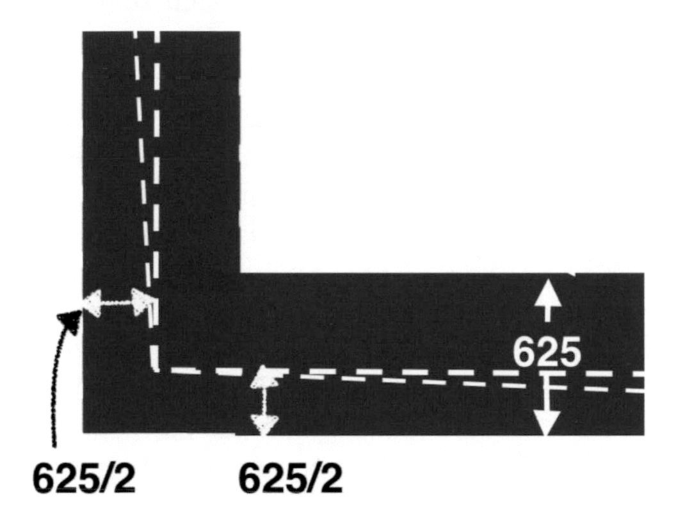

Figure 1.16 Reducing length of trench to the centre line

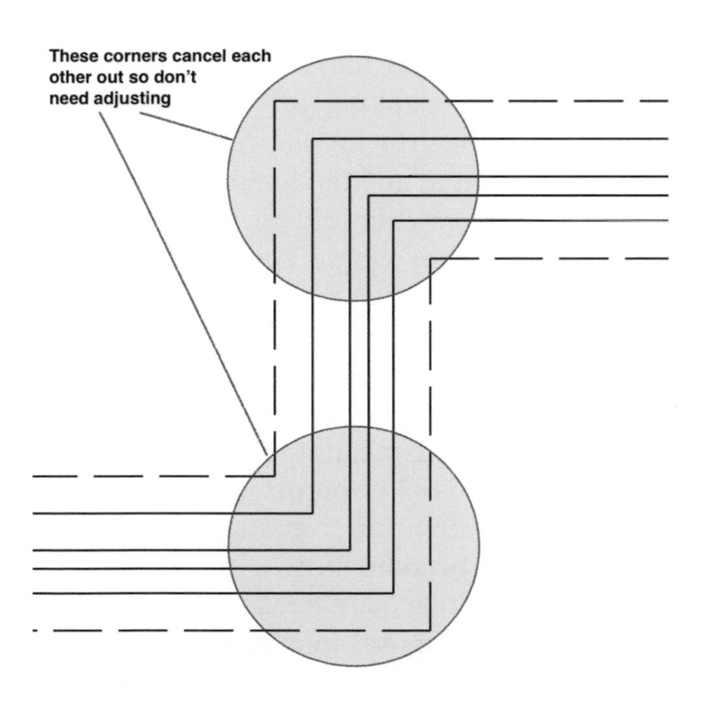

Figure 1.17 Compensating corners cancel out

there are just four corners to adjust, and any corners at regular indented and re-entrant parts can be ignored. The taking-off waste calculation, therefore, shows the overall length (or **girth**) of the building being reduced to the correct length on the centre line by deducting **4 × 2 × 0.5 × 625**, and this gives a figure of **41700mm**.

The length of the party wall is easily determined from figures previously calculated. The perpendicular N–S depth of the building is 8300mm to the outside of the vegetable soil, and this is reduced by the depth of the re-entrant and the width of the external wall trenches at each end, giving a net length of 5550mm.

Calculating the length of the loadbearing partition trenches is a bit more intricate but can be determined by careful observation and analysis. First, the length of the partition itself is based on the dimensions shown on the ground floor plan – 3350mm (between dining and living rooms), 100mm (the thickness of the kitchen/dining room partition) and 2100mm (between kitchen and hall). The last dimension (for the N–S portion at the hall/dining room doorway) is not shown on the drawing, so this has been scaled at 900mm and labelled as *scaled*. This labelling is to alert anyone reading the dimensions that the figure will not be shown on the drawing. Also, in scaling, the taker off will have checked that the drawing runs *true to scale* so that the figure is as nearly accurate as possible. The total length of the partition at 6450mm will be useful later for the masonry above and below ground, but for the length of the trench below, it will need to be shortened slightly to allow for the spread of outside and party wall foundations at each end (i.e., where trench excavation has already been measured).

The spread of the outside wall foundations has previously been encountered (in working out the vegetable soil excavation) as 0.5 × 300mm. The partition trench is therefore reduced in length by 150mm at one end. At the other end, the spread of the party wall foundation is greater, as the party wall is only 250mm thick, but the foundation is 625mm wide. So the "spread" calculation needs to be carried out anew – 625mm foundation minus 250-mm-thick wall = 375mm divided by two = 187.5mm.

Finally, in this set of waste calculations, the taker off has calculated the depth of the excavations. Section A-A shows the depth from ground level to the bottom of the foundations at 1000mm, but 150mm of vegetable soil has previously been stripped. This makes a depth of 850mm and in this case will be the same for all the excavations (in practice, the depth may vary throughout the trenches, as the ground will not be completely flat).

Figure 1.18 shows page AA3 of the taking off with important figures highlighted – the extra lengths on the outside wall girth for the re-entrant, the reduction in length for the corners, the masonry partition length and the length of the partition trench. With all the key dimensions determined as waste calculations, it now remains to dim the items (a colloquial term for writing the items as they would appear in bills of quantities). AA4 shows the item for excavating the trenches and two dimensions only – the 625-mm-wide trench for outer and party walls and the 400-mm-wide trench for the loadbearing partition. The latter is timesed (multiplied) by two, as there are two lengths, one to each house in the block. Exactly the same volume is involved in removing the soil, so the excavation and disposal items are **&** together and bracketed, meaning that the dimensions apply to both items. Descriptions are drawn from NRM2. NRM2 requires that the excavation description give the starting level and the excavation depth in 2-m stages – in this case less than 2m deep. As excavations become deeper, they become more expensive to carry out (the ground may be harder and machines with a longer reach are needed), so NRM2 requires deeper excavations to be identified separately. The disposal item simply states that excavated material is to be removed from site.

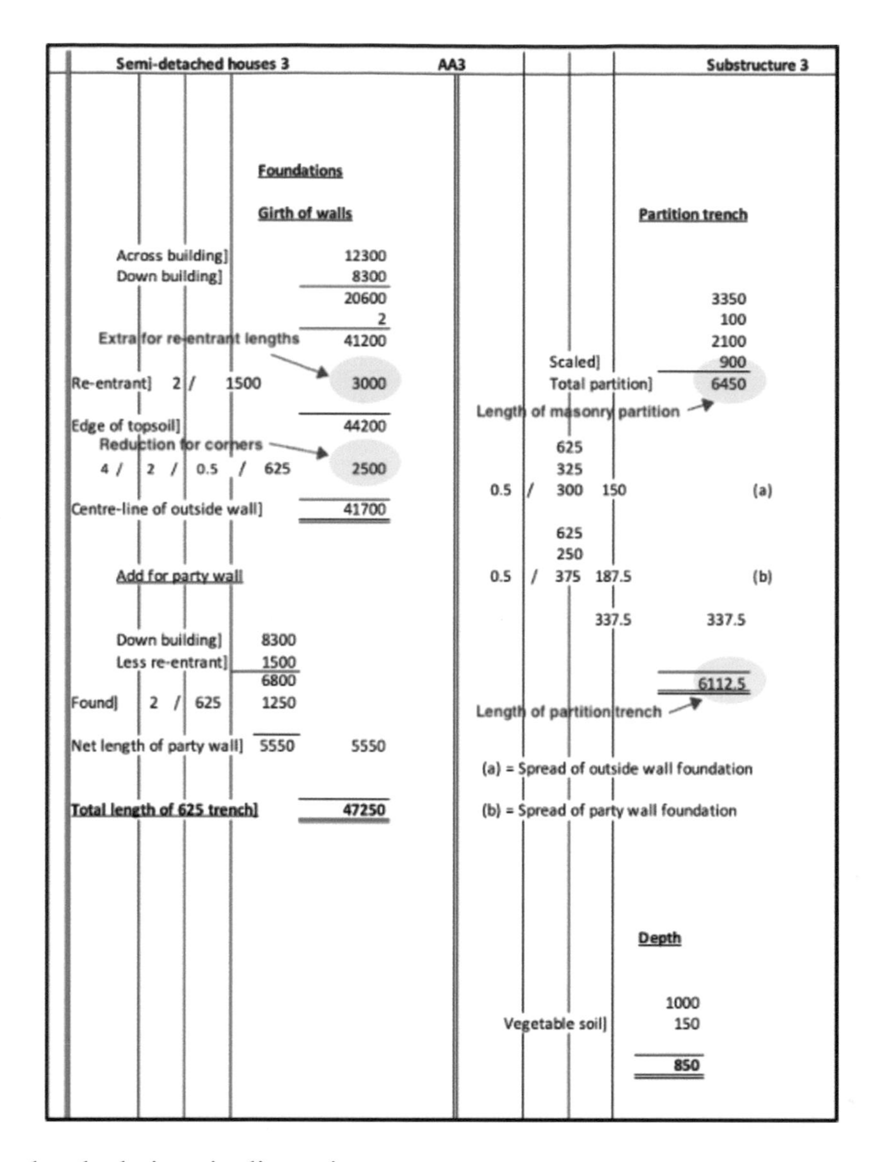

Figure 1.18 Trench calculations in dimensions

Earthwork support

With any excavation it is necessary to consider supporting the face of the earth where the ground has been taken away. For excavations with a sheer face of more than about 1m depth, this will involve physically providing restraint – traditionally timber **planking and strutting** is favoured as it is flexible and gives plenty of warning before failing. However, steel sheet piling is also commonly used. Support needs particular attention for excavations near existing buildings, roadways and other structures, where imposed loads may be high. For shallow excavations, such as trenches for housing, where there is no risk of injury due to collapse, no physical support may be used. However, there will still be a cost in re-excavating and backfilling any earth that falls into the excavation. For excavations where there is plenty of surrounding space, the contractor may choose to ramp back the face – to **over-break** the excavations (make them wider at the top to reduce the risk of collapse). Again, although no physical support will be apparent, over-break excavation and filling has a cost. Where physical support is called for, key variables include the depth of excavation, distance between opposing faces (for narrow trenches it is easier to support opposing faces), whether it is possible to remove the support (and, therefore, re-use it) and whether it is near a road or building.

There are a few items of construction similar to earthwork support, where the method of working determines the physical items involved (**formwork** for concrete and **scaffolding** are similar), and for these the taker off will avoid assuming a method. For earthwork support, NRM2 simply requires that the face of the supported excavation be measured in m^2 and that key variables be identified. What physical support is actually used is left for the contractor and estimator to decide, within the bounds of practicality and safety. AA4 measures one item for support, which includes the trenches and the edge of the vegetable soil excavation. The latter would not normally be measured, as it is less than 250mm high, but it is immediately above the trench excavation, so it is included (in practice, the vegetable soil would probably be excavated beyond the building in any case in grading gardens and paths).

Foundation concrete

The dimensions for the concrete foundations are also shown on AA4. The foundations are the same length and breadth as the trenches, and the depth has been scaled at 400mm. Plain concrete does not contain steel reinforcement and can be placed directly against earth. NRM2 simply requires that horizontal and vertical work in structures be separated.[1] Thicker bulk concrete over 300mm thick is also separated, as it is cheaper per m^3 to form than thinner work. The specification is **performance based**, requiring the work to achieve 15N/mm^2 strength after a period for the concrete to **cure**. The 40-mm aggregate mentioned is the largest stone to be included in the concrete – a fairly "coarse" mix being acceptable for mass foundation concrete.

Substructure masonry

Masonry (normally consisting of concrete blocks and/or clay bricks) is measured in m^2, taking the mean length of the wall multiplied by its height. The lengths of the party and partition walls are easily determined, but working out the length of the outside wall involves more girthing calculations to work out centre lines. AA5 deals with the calculations for the lengths of all components of each wall.

Figure 1.21 shows a corner of the outside cavity wall, consisting of a 100-mm outer skin of concrete blocks, a 75-mm cavity and a 150-mm inner skin also of concrete blocks. The outer skin

Figure 1.19 Mass concrete trench foundation

Figure 1.20 Substructure masonry and suspended beam and block floor

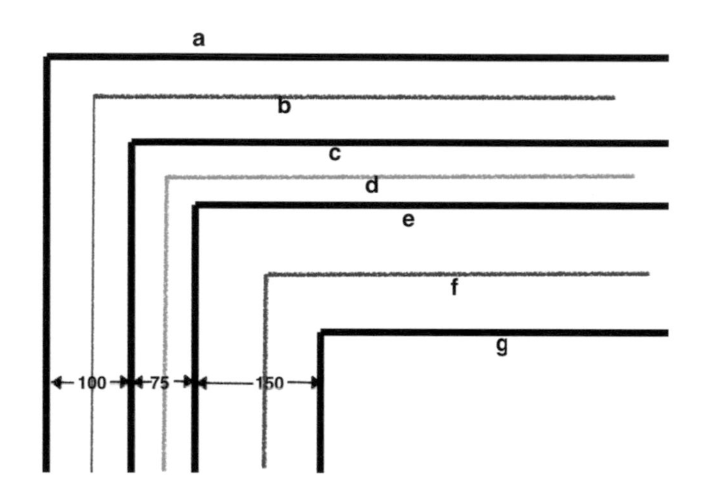

Figure 1.21 Cavity walling

has facing brickwork at the top to give a more attractive finish where the masonry can be seen. The object of cavity walling, above ground level, is to give damp resistance and thermal insulation. Below ground level, these features are not required, and the cavity is filled with concrete.

AA5 starts with a set of girthing calculations, adding initially the overall width and depth of the building and then multiplying the sum by 2 to get 40000mm. The calculation is similar to that for the foundation excavation but based on the face of the wall rather than the edge of the vegetable soil. Adding for the re-entrant (again 1500 × 2), the overall outside length of the wall is 43000mm (**a** in Figure 1.21). There follows a series of girthing calculations using the same method as used for the centre line of the foundation trench (Figure 1.22).

The centre line of the 100-mm outside skin of walling (**b**) is calculated by working out the distance from the outside face to the middle of the skin, multiplying it by 2 for each side of the corner and by 4 to allow for four corners in the building. This is exactly the same method as encountered for the foundations. The resulting 400-mm deduction gives a centre line for the outside skin of 42600mm.

By deducting a further 400mm from this, the inside face of the outside skin is determined at 42200mm (**c**). The centre line of the cavity (**d**) can then be determined in exactly the same way as for the outside skin. The distance to move the line inwards is half the width of the cavity × 2 for

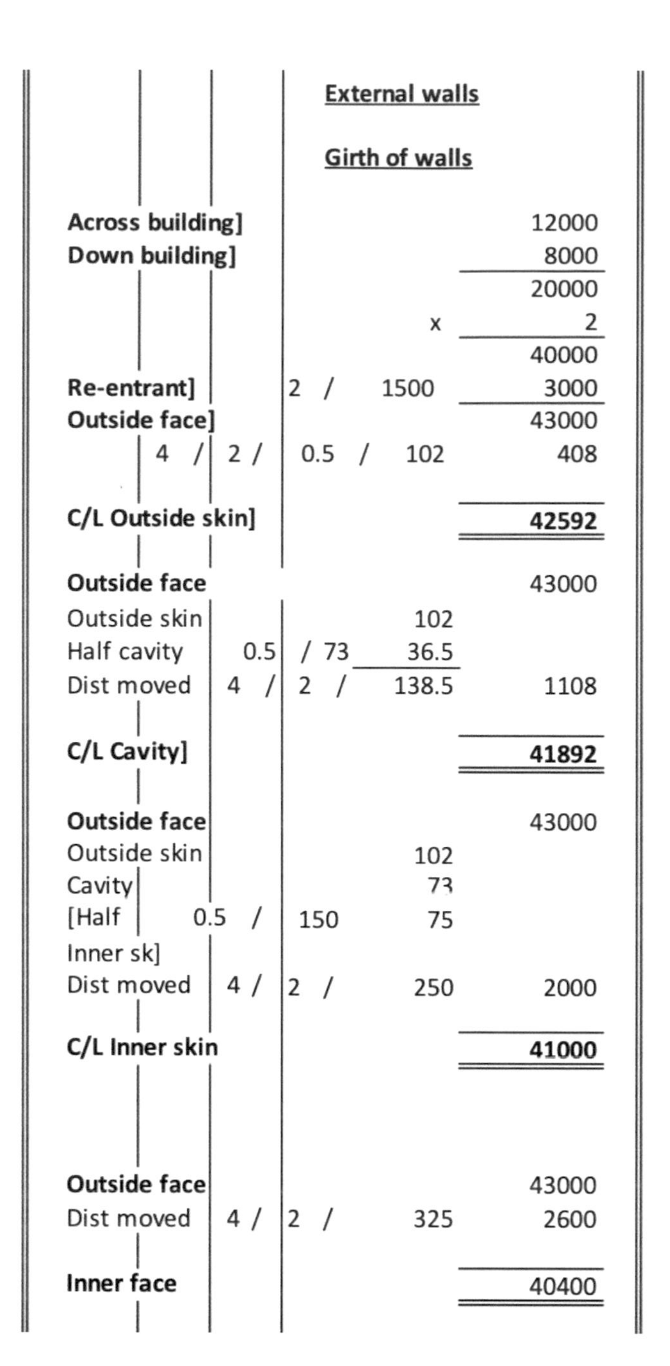

Description	Timesing	Dimension	Waste	Total
			External walls	
			Girth of walls	
Across building]				12000
Down building]				8000
				20000
			x	2
				40000
Re-entrant]	2 /		1500	3000
Outside face]				43000
	4 / 2 /	0.5 /	102	408
C/L Outside skin]				**42592**
Outside face				43000
Outside skin			102	
Half cavity	0.5	/ 73	36.5	
Dist moved	4 /	2 /	138.5	1108
C/L Cavity]				**41892**
Outside face				43000
Outside skin			102	
Cavity			73	
[Half	0.5 /	150	75	
Inner sk]				
Dist moved	4 /	2 /	250	2000
C/L Inner skin				**41000**
Outside face				43000
Dist moved	4 /	2 /	325	2600
Inner face				40400

Figure 1.22 Girth of external walling items

both sides of the corner × 4 for the four corners. The total deduction is 300mm, and the length of the cavity is 41900mm. Reduce this length by 300mm, and the outside length of the inner skin is 41600mm (**e**). Repeat the centre line calculation for the 150mm inner skin (deduct 4 × 2 × 0.5 × 150 = 600mm), and the centre line of the inner skin is 41000mm (**f**). A further deduction of 600mm gives the length of the inside face of the wall at 40400mm (**g**). The last figure is not needed for the walling but will be useful for other calculations (such as the internal finishings).

AA5 also shows the calculations for the length of the party wall (Figure 1.23), starting with the overall perpendicular N–S depth of the building of 8000mm, deducting the re-entrant of 1500mm and the thickness of the outside wall at each end. The party wall cavity is slightly longer than the block skins of walling, as the cavity extends into the outside wall inner skin, as shown in Figure 1.24.

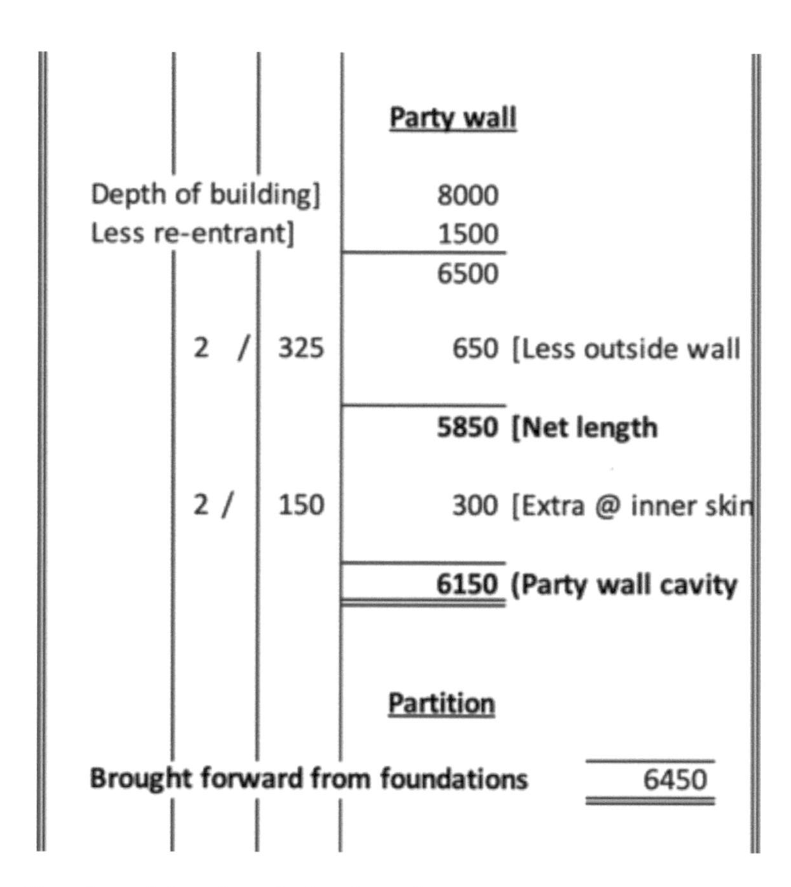

			Party wall
Depth of building]			8000
Less re-entrant]			1500
			6500
2 /	325		650 [Less outside wall
			5850 [Net length
2 /	150		300 [Extra @ inner skin
			6150 (Party wall cavity
			Partition
Brought forward from foundations			6450

Figure 1.23 Party wall and partition foundation length

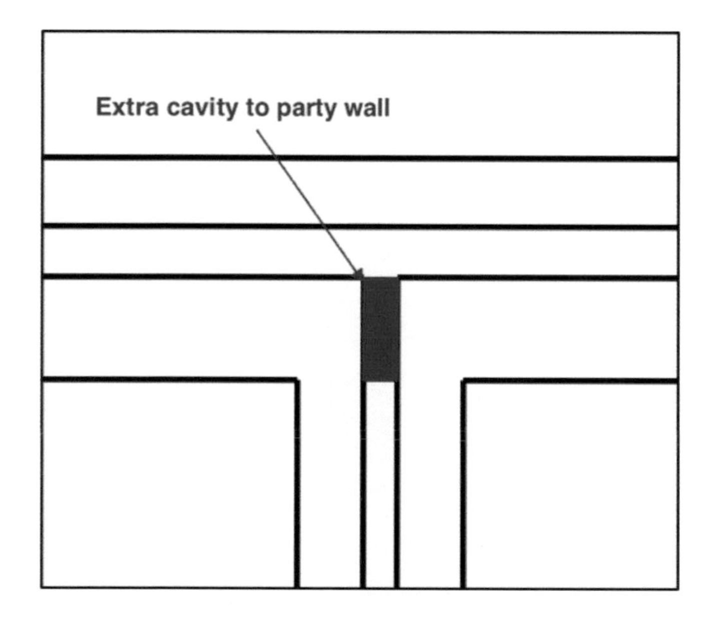

Figure 1.24 Extra length to party wall cavity

The length of the loadbearing partition has previously been calculated, with the trench, as 6450mm, so it is not necessary to re-calculate this length. The figure is simply brought forward and written down as a reminder to those following the dimensions.

The depths of the walls are all the same and can be calculated by taking the depth from underside of foundations to the ground, adding on the height from ground to damp-proof course and deducting the thickness of the foundation. Note the order of working – carry out the additions first and then make any deductions. Always over measure first if possible and then reduce figures afterwards.

Having calculated the lengths and depths of all the substructure walling, it remains, in AA6, simply to enter the dims. First, the 100-mm skins of hollow walls are entered – these include both the outer skin to the outside wall and the skins to the party wall (multiplied by 2). The description follows NRM2 at 14.1.2.1.1, and the specification follows the brief specification provided. The distinguishing features of the specification are that it is for dense concrete blocks using a cement mortar. The fairly high-strength block (10.5N/mm^2) is specified to give sulphate resistance in acidic ground rather than for strength, and cement mortar (without lime or plasticiser) is used for the same reason – more related to durability than compressive strength. Blockwork is normally laid in "**stretcher bond**" (that is, with the longest "stretcher" face showing on the whole wall surface). However, occasionally blocks will be specified as being *laid flat* (i.e., on the stretcher face), particularly for solid party walls. The bond of the blockwork affects the price of the work, hence the NRM2 requirement for stating it.

The 150-mm inner skin of blocks is of the same specification as the outer skin and can be simply written as "ditto", with the correct dimensions. The "ditto" applies to the whole of the previous item except the thickness of 150mm. The loadbearing partitions are not in skins of hollow walls, and NRM2 requires simple walls to be measured separately (14.1.2.*.1). The * simply means that there is no descriptor at that level of description.

Features to note in the dimensions are:

1 The use of **signposts** ("outer", "party") to show what the dimensions refer to. This is of great help in following taking off if changes are subsequently needed.
2 The party wall skins of hollow walls are timesed by 2 to allow for both skins.
3 The loadbearing partitions are timesed by 2, as there are two dwellings in the block.

The wall cavities

Forming the cavities between skins of walling involves little cost except for the wall ties, and the specification for the ties is given in the description. NRM2 requires that each width be given separately, so there are two items, one 75mm wide for the outer wall and the other 50mm wide for the party wall.

Figure 1.25 Cavity walling in substructures

The specification requires that the cavity be filled with concrete, which is measured separately. It is shown to be up to 150mm below the damp-proof course. The side calculation above the filling item, therefore, shows the brickwork depth less 150mm to give the height of the filling. NRM2 11.5.1.1 simply requires that the filling be measured in m^3 and described as "vertical work" less than 300mm thick. A fine stone aggregate is specified in order to fit into the narrow space.

Facing bricks

AA8 makes an **adjustment** for using facing bricks for the top four courses of the external brickwork to the outside wall. The architect has provided a detailed specification for the bricks and their treatment. The specification requires that four courses be facings, and as one course of bricks is 75mm high (allowing for a 10-mm mortar joint), 300mm of facings is therefore required to the same length as the outer skin of the outer wall.

The description follows NRM2 14.1.1.1.1 and includes an outline of the specification. Key determinants of both include the thickness of the wall, described as "half brick thick". Traditionally bricklayers work by thickness of their key material in "bricks" rather than measured dimensions. Describing the exact thickness of a brick (102.5mm) also introduces a degree of spurious accuracy. Wall thicknesses are based on the length of a standard brick including a 10-mm mortar joint – 225mm long, 112.5mm wide and 75mm high:

- One brick thick – 225mm – the length of a brick
- Half brick thick – 112.5mm – the length of a brick divided by two.
- One-and-a-half brick thick – 337.5mm – used occasionally in historic repair and restoration.
- Two brick thick – 450mm – again used occasionally for historic work.
- And so on!

The actual thickness of the wall is important to designers and constructors, so exact dimensions may be specified. This can be difficult, as some walls (for example, a one-brick wall) will vary in exact thickness depending on the direction of bond of the bricks in the wall.

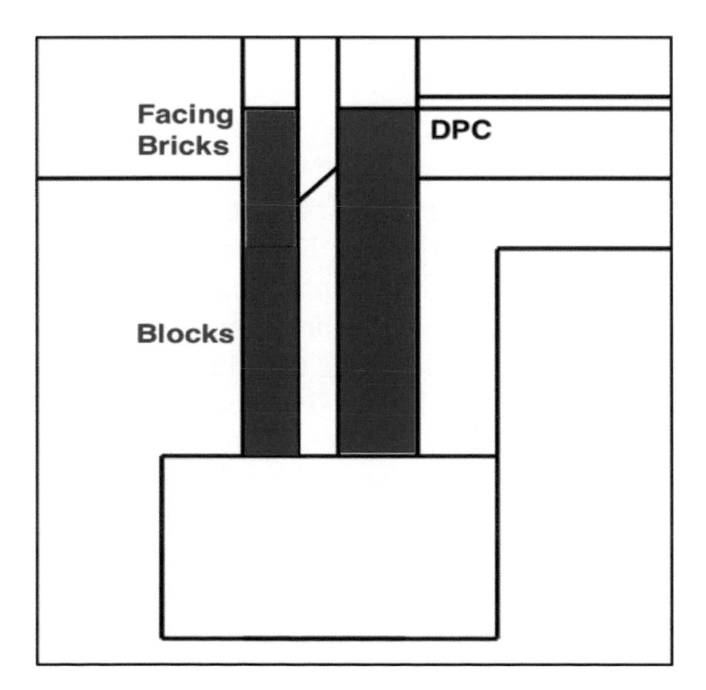

Figure 1.26 Facing bricks to external skin of wall

Another determinant is the treatment of the face of the wall. Treatment may be to one face (as in our example) or to both (for example, to a garden wall or the wall of a garage where the inside face will be visible). **Facing** a wall does not relate to the specification of the brick but to the work in selecting a sound, clean brick, showing the best face outwards and bringing the bricks to a neat, level line. **Pointing** the wall refers to treating the mortar joint between the bricks. If the work is not to be seen, the mortar will usually be **left as laid**, with no further treatment, or raked back to give a **key** for plaster. For visible work, a simple rounded "bucket handle" or "flush" joint formed at the same time as laying the bricks (described as "as the work proceeds") is an economical solution. More expensive solutions involve laying the bricks and raking out the joints, then coming back and pointing up with a different mortar. In our example, a "weatherstruck" joint is specified, pointed as the work proceeds. For detail on the intricacies of brickwork and pointing, you need to refer to specialist texts.

The type of brick to be used is, in this case, specified by the manufacturer, and the bedding is the same as for the blocks – a strong cement mortar. The bond (i.e., the method of interleaving the bricks) is described as stretcher bond, the same as the blocks. That is, the visible face will all be of the *stretcher face* of the brick, nominally 225 × 75mm, including the mortar joint, staggered to give the characteristic brick pattern. There are a large number of bonds used for brickwork, but for half-brick walls, stretcher bond is nearly always used, as it minimises cutting the bricks. Again, reference to specialised texts is necessary for further detail.

The facing brick item is followed by an adjustment, involving *deducting* the same area for the concrete block outer skin previously measured. The deduction is bracketed and **&** on to the facings item, so the dimensions apply to both. The two items also illustrate a basic principle of measurement previously encountered when measuring the vegetable soil excavation. For the vegetable soil, the area was measured overall with an adjustment for the unwanted part at the entrance. For the facing bricks, the bricks were measured first with an adjustment for the concrete blocks. In both instances, the process involved *over-measurement* first and then adjustment downwards. That way, forgetting the adjustment will lead to too much rather than too little quantity. This is a better solution than the other way round – when the quantities are corrected, it will involve a saving rather than an extra.

Damp-proof course

Finally, on AA8, the **damp-proof courses** (**DPC**s) are measured. These are actually laid as an early superstructure item but traditionally measured with the substructure. DPCs are thin strips of impervious material such as vinyl bedded in mortar on the brickwork. For cavity walls, there is a separate DPC for each skin (not a wide strip over the whole wall). There is not much difference in cost for supplying and laying any DPCs less than 300mm wide, so NRM2 simply requires that these be collectively identified and measured in linear metres (NRM2 14.16.1.3). The quantities involved are, therefore, the total lengths of the two skins to the outside and party walls and the loadbearing partitions. The DPCs are described as being horizontal (as opposed to vertical, **raking**, etc.), but no allowance is made by the taker off for laps in the length or at corners. The estimator is required to make such allowance in pricing the item.

Backfilling adjustments

AA9 deals with **backfilling** adjustments. At the sides of all the substructure walls and loadbearing partitions, between the foundation concrete and the ground or oversite there is a volume of excavated trench which needs to be filled in after the foundations have been installed. The filling to the outside walls of the building can be in earth arising from the excavations (provided that the ground is for a garden or similar use). The filling to the inside of all walls and partitions must be in hardcore for this type of ground-bearing construction. Working out the length of the filling follows similar principles to those used for the trenches and brickwork. On AA9 the girth of the outside earth filling is calculated first, by taking the previously calculated girth of the outside wall (43000) and increasing this by using the now-familiar girthing calculations of 4 × 2 × 0.5 × 150mm (= 600mm). The 150mm is the spread of the foundation, as previously identified when the vegetable soil excavation was measured (marked in red in Figure 1.8).

The girth of the filling to the inside of the outside walls is similarly calculated, by taking the previously calculated girth of the inside of the wall (40400) and decreasing this by the same 600mm to give 39800mm. The lengths of foundations to party walls and partitions have both been previously calculated, and the respective figures can be brought forward. The width of the filling is the same as the spread of the foundations – summarised as 150mm to the outside walls, 187.5mm to the party wall and 150mm to the partitions. The depth of the filling in all cases is 450mm – the total depth from ground level to bottom of foundations, less the foundation and less the vegetable soil.

By careful observation, you will notice that the filling to the inside of the outside walls has been measured over the ends of the party walls and partitions. Similarly, the filling to the party wall has been measured over the ends of the partitions. This is illustrated in Figures 1.28, 1.29 and 1.30, and a deduction is made for the blue area.

These small adjustments are not likely to make a significant difference to the net measured quantities of the item, and an experienced taker off, exercising judgement, might well ignore them. In training, however, all such adjustments should be made to show that they have been identified. Another way of putting this is that short-cuts in measurement can be

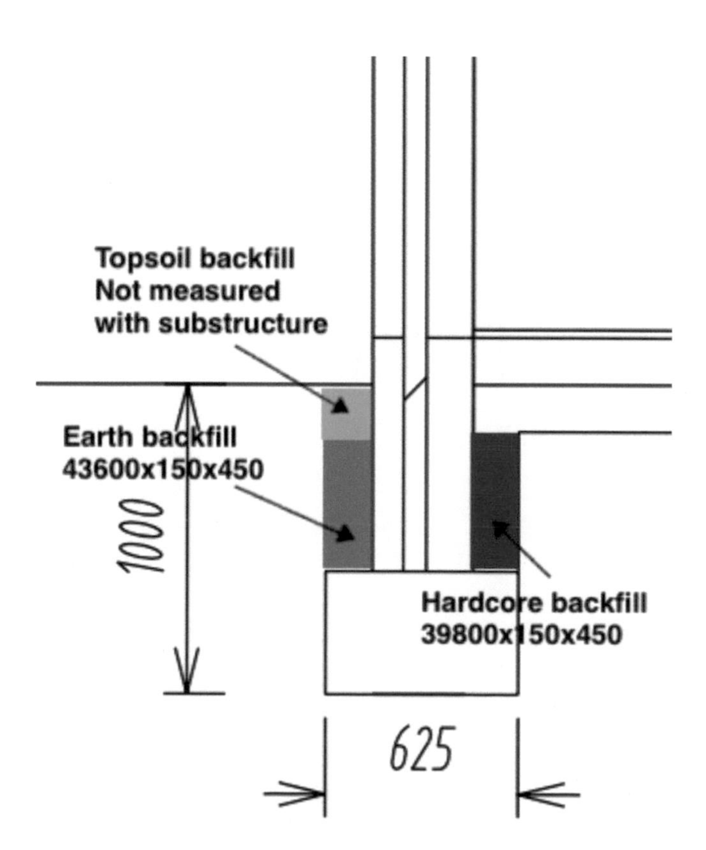

Figure 1.27 Backfilling to external walls

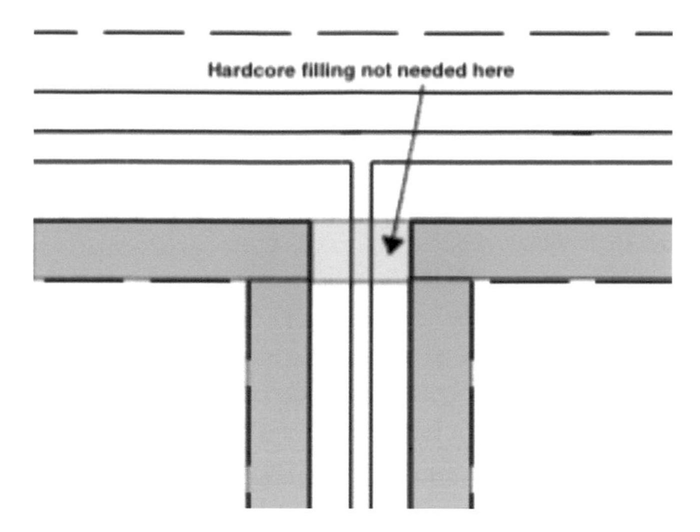

Figure 1.28 Deduction to trench backfill at party wall

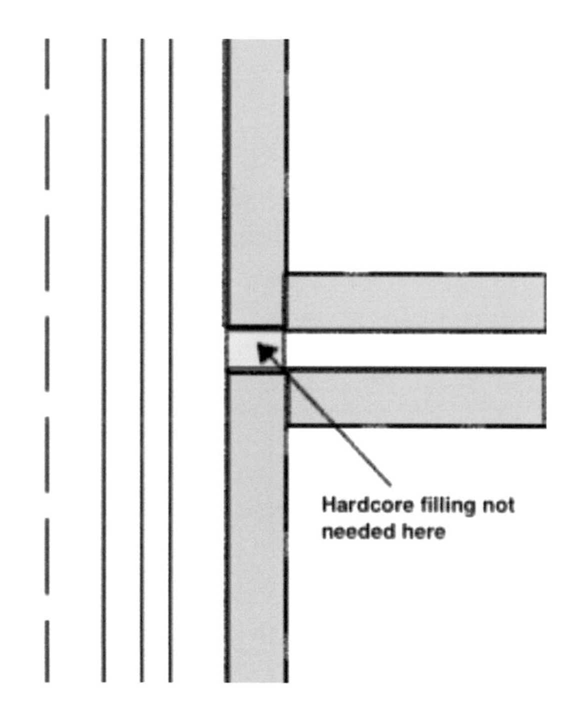

Figure 1.29 Deduction to trench backfill at partition

made, but only if the taker off has sufficient knowledge and experience to put their effect into context.

Finally, note the comment at the bottom of AA10 that no adjustment has been made for returning topsoil (vegetable soil) back to the outside walls. This is shown green in Figure 1.27 but marked as not measured in the substructure. Not measuring the topsoil backfilling is based on the practical assumption that treating topsoil, grading garden areas, constructing garden paths and so on will be measured in a separate *External Works* section of the bills of quantities, possibly by other takers off. The note is to remind the taker off and those following the dimensions of this assumption.

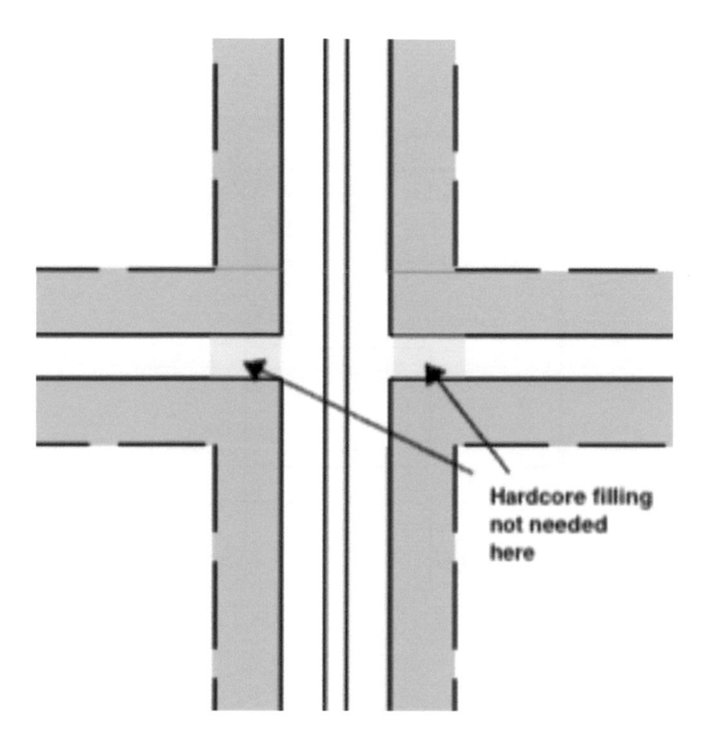

Figure 1.30 Deduction to trench backfill to party wall at partition

Oversite construction

The design for the floor uses a traditional "**ground bearing**" concrete bed laid between the walls and loadbearing partitions over a hardcore sub-base with a polythene damp-proof membrane laid between. The polythene is **dressed** up the walls next to the concrete and into the edge of the damp-proof course in order to ensure there is no path for dampness from the ground. As the concrete is not tensioned at any point, there is no requirement for **steel reinforcement** (rebar). Sometimes, however, light **mesh steel reinforcement** is specified to account for any slight unevenness in the ground. With ground bearing beds, it is essential that all backfilling to trenches be in hardcore or similar granular material; otherwise settlement is possible. Ground bearing beds are less common now (although still current), and suspended concrete floors (as shown in Figure 1.20) are an alternative. There is also a requirement for insulation, and this may be placed between the concrete and the hardcore base. In this example however, insulation is placed over the concrete and is, therefore, not measured with the substructure but with the finishings.

AA11 shows the calculations for the area of the oversite using the same principle as for the area of vegetable soil. The overall footprint of the building is 12000 × 8000mm, and the outside walls are 325mm thick. Taking two thicknesses away in each direction therefore gives internal dimensions of 11350 × 7350mm. The deduction for the re-entrant at the entrance area is 1500mm in the N–S direction, but in the E–W direction, the 4000-mm width previously calculated is increased by the thickness of the walls.

NRM2 requires that hardcore beds less than 500mm thick be measured in m^2 stating the thickness (NRM2 5.12.2.1.1). Beds of hardcore, concrete and similar material may be:

- Laid level, as is usual for floor construction.
- Laid sloping – that is, the bed is of constant thickness, but the supporting structure or earth also slopes. This might be the case for a roadway or path.
- Laid to falls – that is, the bed is thinner at one edge than the other, usually to allow water to run off to the edge. This might be the case for a parking area or flat roof.

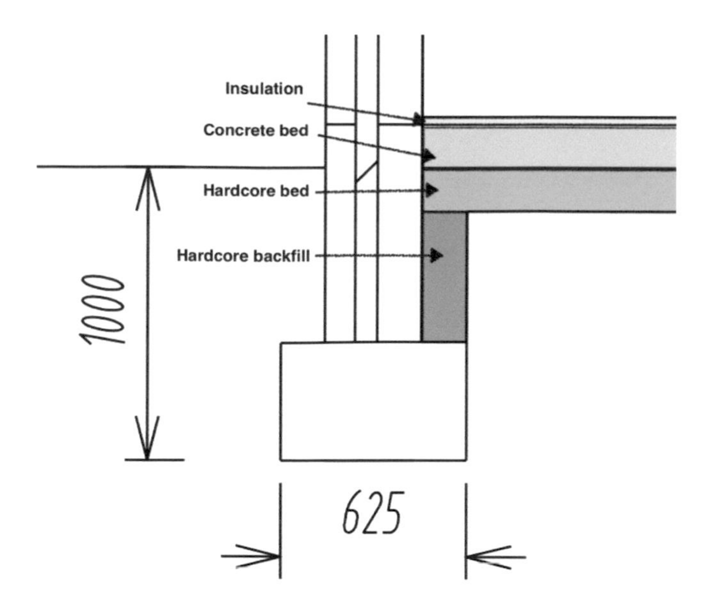

Figure 1.31 Oversite construction

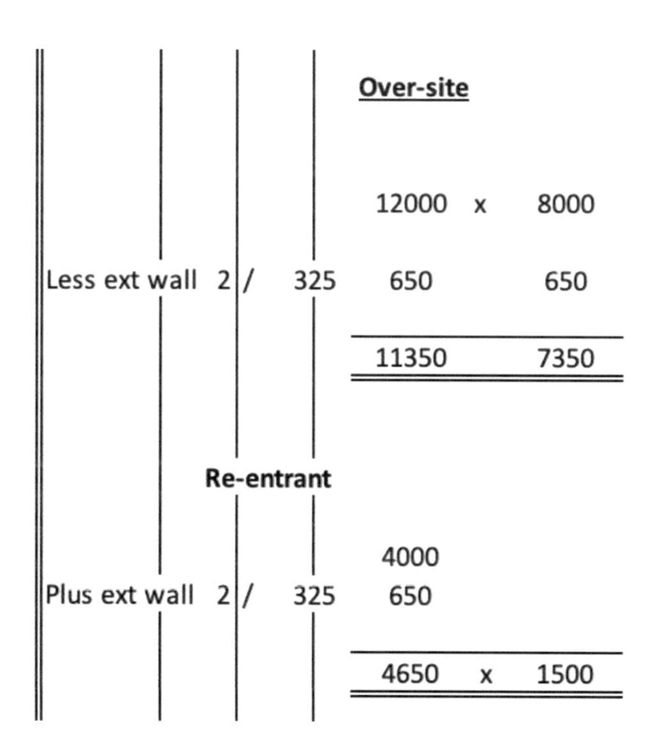

Figure 1.32 Oversite waste calculations

- Laid to falls and crossfalls – that is, the bed is thinner at two adjacent edges and thicker at the opposite edges to allow water to run off to a single point. Also common for parking areas and roofs.
- Laid to camber – that is, the bed is thicker in the centre but falls away to two opposite outside edges. This might also be the case for a roadway or path.

As all of these variants have a different costs, NRM2 requires that each be measured separately and that they be stated in the description of the hardcore bed. The treatment to the surface of the hardcore (involving **blinding** with sand to give a smooth finish to receive the polythene) is not

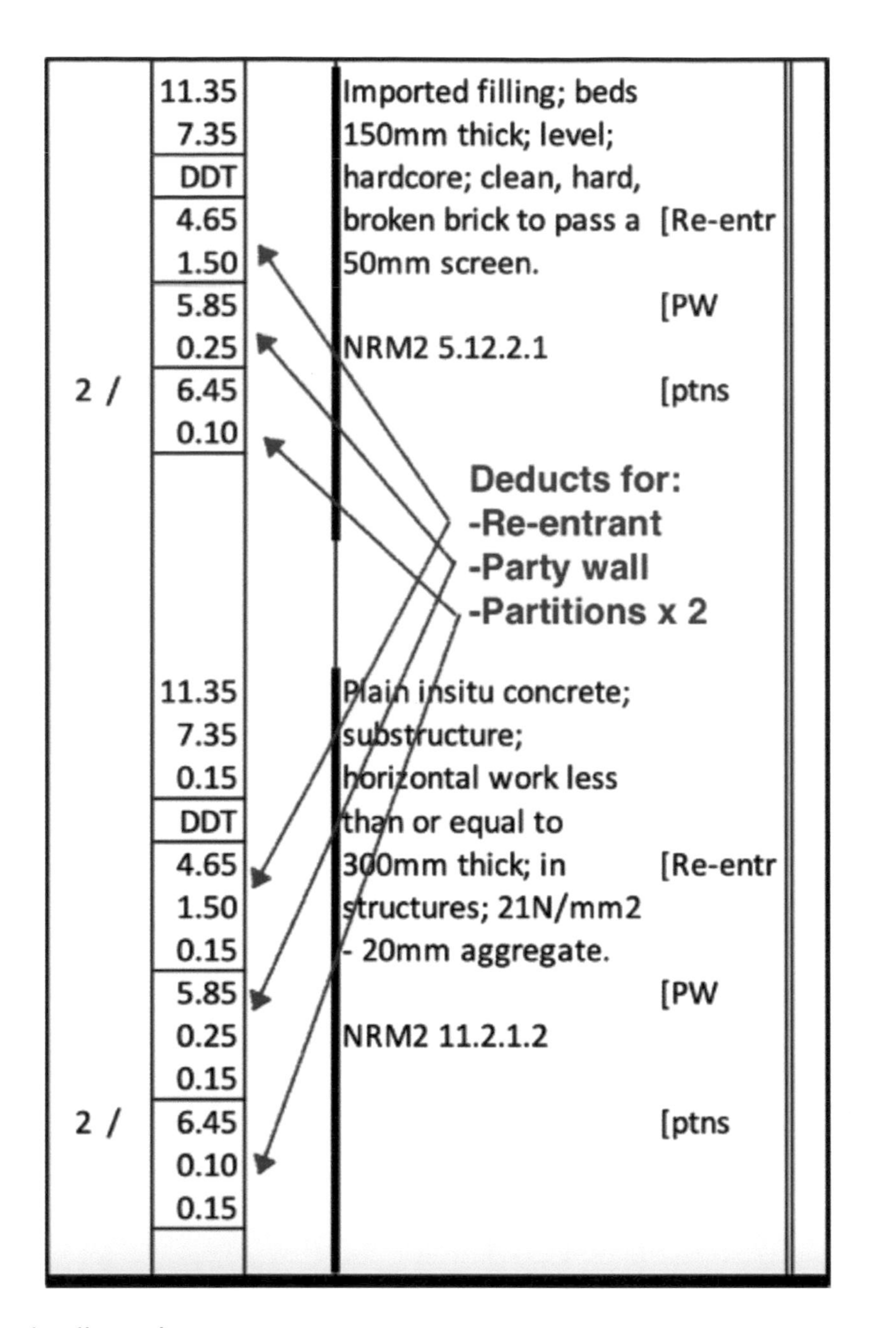

Figure 1.33 Oversite dimensions

measured separately but deemed included in NRM2. This means that the estimator must make any allowance for the cost of blinding.

The concrete bed is measured in m^3 stating that the maximum thickness of the pour is less than 300mm (NRM2 11.2.1.2) and whether the item is vertical or horizontal. Treating the surface of the concrete, which in this case is by tamping with a machine, is measured separately (NRM2 11.12.1). As with hardcore, surface treatments to form falls, crossfalls or sloping are given separately.

Finally, the polythene damp-proof membrane, provided it is over 500mm wide, is measured in m^2 (NRM2 5.16.2.1). Although not expressly required by NRM2, the size of any lap in the membrane is specified and itemised in the description, but no allowance is made in measurement for the laps. The membrane is dressed 150mm up the sides of all walls and connected to the damp-proof course, so the area of this extra has been included in the dimensions for the item. The length at the outside walls is the previously calculated girth of the inner face of the wall (40400mm),

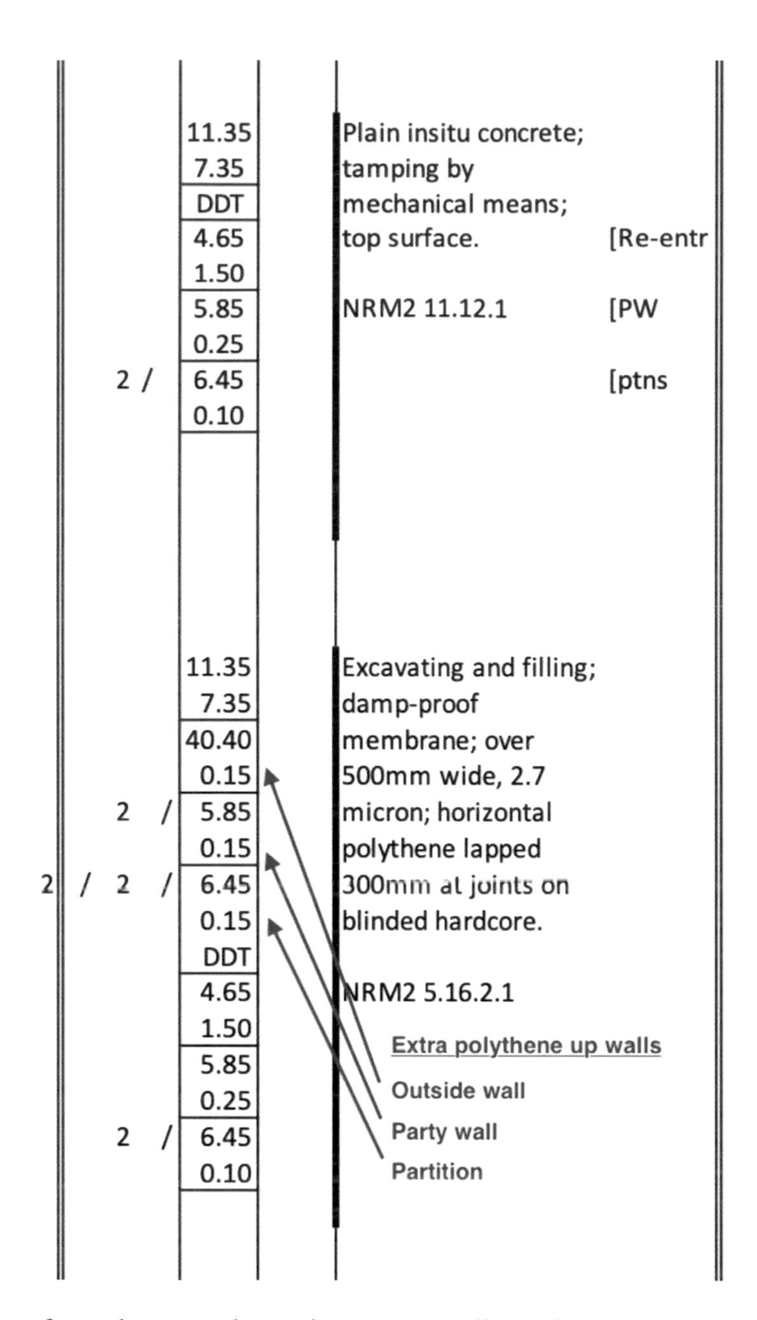

Figure 1.34 Damp proof membrane and tamping concrete dimensions

and the lengths at the party wall and partition are as previously calculated, multiplied by 2 for both sides of the wall. Lengths are all multiplied by the height of the upturn, 150mm. Unlike the hardcore trench filling and because the membrane is a very low-cost item, no deductions have been made to the lengths of the upturn for the party wall and partition **abutments**. However, as with the other items (hardcore and concrete), there is a deduction for the party wall and partitions where these protrude through the oversite.

Substructure – summary

Learning taking off by starting with the substructure of a building is challenging. It involves introducing fairly complicated technology at the same time as grappling with the skills and techniques of taking off building work. Nevertheless, there is a logic in starting at the point where construction starts and working upwards towards the finish. By carefully working through the example and following this by tackling further exercises, proficiency will quickly be gained. What has been covered in this section is the technology and procedure for taking off substructures but more importantly also the form and thought process involved in all taking off. In this respect, learning taking off is similar to any other skill, including both manual trades and professions. At first, any skill is extremely tricky, but by careful application, practice and knowledge of results, expertise is obtained. Expect to spend much time "in the ground", puzzling over what has been measured, but don't be too surprised when the approach to all other sections involves employing the same analytical logic and dexterity rapidly improves.

Key points covered in this chapter:

- Outline of the technology of domestic substructures
- The sequence of construction activities for domestic substructures
- Preliminary activities for taking off
    - Reading and handling drawings
    - Query sheets and their use
    - Identification and referencing of taking-off output
    - Taking-off lists
    - Dimension checks
- Calculating areas for simple buildings for excavation and oversite
- Calculating centre-line girths for trenches, concrete foundations, masonry and so on
- Dealing with re-entrants in a building for area and length calculations
- Identifying items for taking off substructure
- Constructing item descriptions for domestic substructures using NRM2
- Setting out taking off using accepted conventions
- Applying explanations to the example

Note

1 It also requires that all work in substructure be separate, so the concrete and all other work is in a separate substructure section in the bills of quantities.

2 External walling, party walls and partitions

In our project, walls and some partitions are of loadbearing masonry. External walls are of cavity construction. Cavity walling was initially adopted to limit the penetration of damp but adapted to improve thermal performance. Cavities are now filled, fully or partially, with insulant, and thermally efficient lightweight concrete blocks are used for the inner skin. The **party wall** between the two dwellings is of dense concrete blocks in order to support the roof construction and provide fire and sound resistance. It is taken up to the underside of the roof coverings in order to ensure separation of the dwellings and maintain fire resistance in the loft.

Some internal partitions support the first floor and roof and are taken down to foundations, so they need to be of a suitable loadbearing strength. In our project, the **loadbearing partitions** are of the same dense concrete blocks as used for the party wall. Other partitions simply divide spaces into rooms and, subject to being stable and sound proof, can be of lighter construction. For modern housing, plasterboard on a lightweight framing is common, but, in our project, the **non-loadbearing partitions** are of lightweight blockwork.

Masonry requires support over windows and doors, and for modern housing this is usually provided by purpose-made steel cavity wall lintels. For our project, over the ground floor openings, lintels combining the function of supporting masonry and providing water resistance are specified (Figure 2.1). There is no masonry over the first-floor windows, and the steel lintels are a lighter type suitable for supporting the roof construction and roof tiles (Figure 2.2). Support is also required over internal openings in loadbearing and non-loadbearing partitions, and, as concrete blockwork is used for both, a simple pre-stressed precast concrete lintel is specified.

The **jambs** (sides) of external openings require some treatment to close cavities, provide a base for the plaster **reveals** and ensure that moisture will not be transmitted to the interior. A modern effective way of doing this is to use a simple purpose-made profiled plastic **cavity closer** (Figure 2.3). At the base of windows and doors, it is also common to close the cavity at the sill, either with a similar cavity closer or by returning the blockwork inner skin over the cavity. In our project, the cavities are left open at the sills and closed by the window boards fitted as part of the **first fix** joinery activity and measured with the windows and doors.

Preliminary items

AA13–AA23 cover the taking off for the external walling, party walls and partitions. Preliminary referencing and labelling are dealt with in the same way as the substructure, including referencing the taking-off drawing numbers and revisions. A detailed taking-off list is as shown in Table 2.1.

After listing the drawings used for this section and constructing a taking-off list, a dimension check is again carried out. If one taker off were measuring the whole project, just one detailed

DOI: 12.01/9781003253129-3

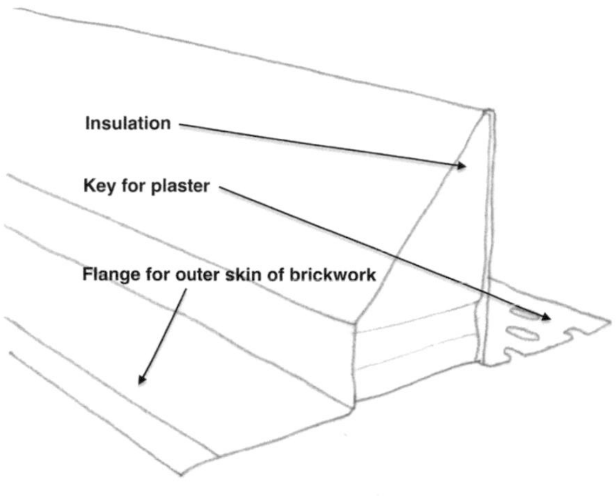

Figure 2.1 Steel cavity wall lintel

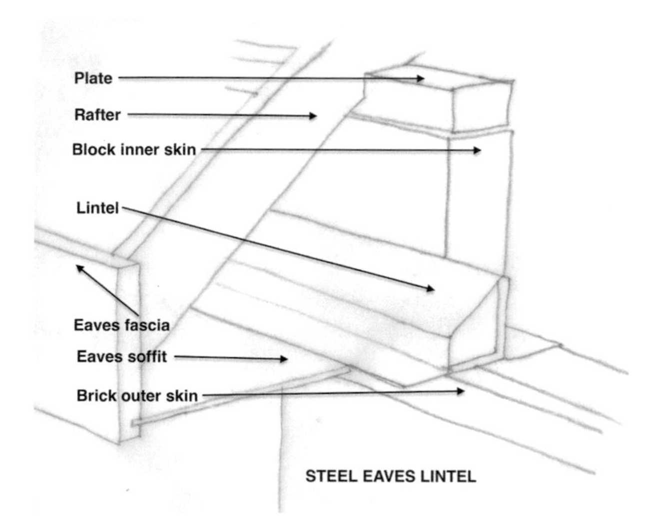

Figure 2.2 Steel eaves lintel

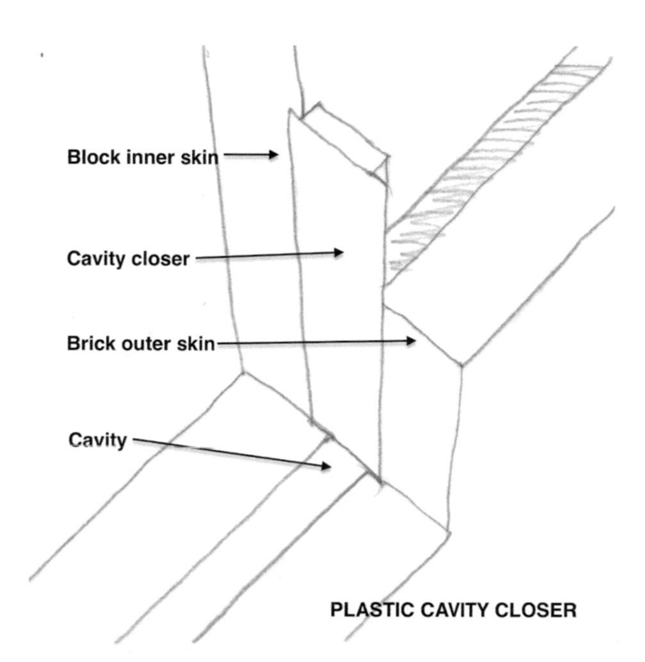

Figure 2.3 Plastic cavity closer

Table 2.1 Walling and partitions taking-off list

Item	*Work involved*
Facing bricks	The outer skin of the external cavity wall involving a facing brick, pointed and faced up as visible brickwork. The facing bricks are slightly thicker at 102mm than the blocks used below ground level. In order to maintain a constant wall thickness of 325mm, the designer has reduced the width of the cavity from 75mm below ground to 73mm above ground.
Cavity and insulation	Forming the cavity in much the same way as below ground; however, the specification includes providing insulation, and this item is measured separately.
Inner block skin	As with the substructure inner skin, this is 150mm thick but formed with insulating blocks. A detailed specification would be provided to back up the brief details included with the drawings.
Party wall	As with the substructure, the party wall is of two skins of 100-mm dense concrete blocks but using a mortar with a different composition. Dense concrete is used, as thermal insulating qualities are not needed and to improve sound insulation.
Party wall cavity	Forming the cavity in the same way as with the substructure. Insulation is not needed for the party wall.
Loadbearing partitions	The type of construction used for the two houses requires that some internal (loadbearing) partitions take loads from the floor and roof and transfer these to the foundations. The substructure loadbearing block partitions have previously been measured, and the specification is similar above ground. As the specification is different from non-loadbearing partitions (dense as opposed to lightweight concrete block construction), they need to be measured separately. Partition heights are also different, as loadbearing partitions pass through the first floor up to the upper floor ceiling. Loadbearing partitions, as they give support to the floor and roof structure, are constructed as the main structure is raised.
Non-loadbearing partitions	Non-loadbearing partitions are often of lighter materials, including timber "**stud**" and plasterboard. They are usually erected after the main structure and ceilings are installed. In this building, non-loadbearing partitions are made from lightweight blocks.
Adjustments for external openings	All the external walling, party wall and partitions have been measured overall without deduction for the doors and windows.
Deduct facings	The area of external doors and windows is deducted out. As a check that all openings have been adjusted, it is good practice to add up the number of openings measured and compare this with the number counted on the drawings.
Deduct cavity and insulation	Same treatment.
Deduct inner block skin	Same treatment.
Cavity closer	In measuring any superficial item (an area of brickwork, opening, flooring, roofing, etc.) it is good training to immediately consider the treatment of the surrounding perimeter. In this case, deducting the window/door area will lead to work to the head (lintel, arch or similar feature), the sill (brick, stone, tiled sill or similar) and the jambs. The construction of our houses involves simple steel lintels over the opening, plastic cavity closers to the jambs and no treatment to the sills. The simple plastic cavity closers in this building are measured in linear metres.
Lintels	The steel lintels used in this building are measured and purchased as enumerated units. Various lengths of lintel are used of two types, one for the cavity walling on the ground floor and the second under the eaves on the first floor.
Adjustments for internal openings	The same principle as adjusting external openings but much simpler.
Deduct blocks	The relevant areas are deducted for internal door openings.
Lintels	For internal doors, there is no treatment at the perimeter except for providing concrete lintels over the openings. Other treatment, such as frames, door linings, thresholds and so on, would be measured with the finishings or windows/doors and not with this section.

check would be carried out at the start of all taking off. However, assuming each section is measured by a different individual, it is prudent for each to make his or her own checks. The dimensions checked involve the overall E–W and N–S dimensions of 12000 and 8000, but the detailed dimensions are, this time, taken off the first-floor plan.

External walls

AA14 deals with the waste calculations for the lengths and heights of the components of the external walls. A slightly different method has been used for girthing here from the superstructures to illustrate that there is more than one way of getting the right answer.

The girthing calculation starts, as with the substructure, by adding the overall breadth and depth of the building of 12000 × 8000mm and multiplying the answer by two. It adds 2 × 1500mm for the re-entrant at the entrance area, giving 43000mm overall for the outside girth. For the centre line of the outside skin (represented by the red line in Figure 2.5), the distance moved in is half the width (102mm) of a standard brick. This is slightly wider than the 100mm block outer skin in

Figure 2.4 Insulant-filled cavity walling

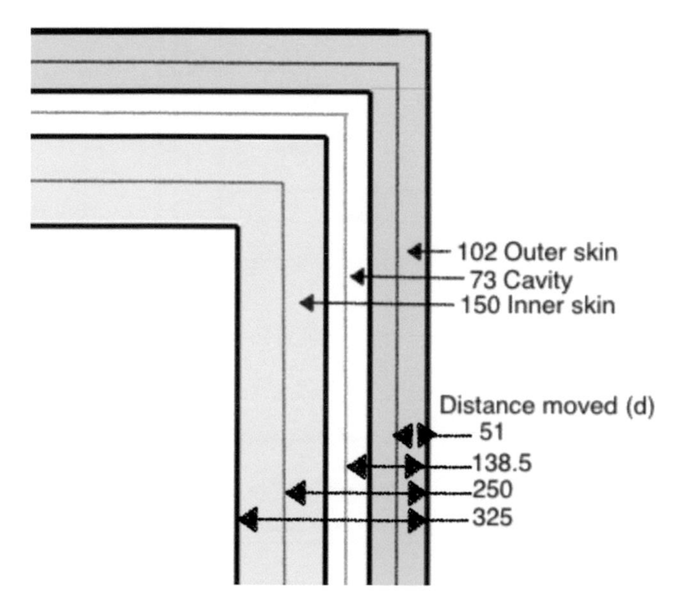

Figure 2.5 Superstructure cavity walling

the substructure. However, the overall width of the wall remains the same as in the substructure at 325mm, as the architect has specified a narrower 73-mm cavity above ground. The girthing calculation is 4 × 2 × 0.5 × 102mm = 408mm deducted from the outside girth, giving 42592mm as the centre line of the outside skin.

For the centre line of the cavity (green in Figure 2.5), the calculations return to the outside girth of 43000mm, and the distance moved is calculated as being the width of the outside skin and half the width of the cavity – that is, 102 + 73 ÷ 2 = 138.5mm. The girthing calculation is the same as before, but the actual distance moved has been calculated (i.e., not the whole width of the item multiplied by 0.50). The calculation is 4 × 2 × 138.5 = 1108mm deducted *from the outside girth*, giving 41892mm. For the centre line of the inner skin (blue in Figure 2.5), the distance moved is the width of the outside skin plus cavity plus half the width of the inner skin – that is, 102 + 73 + 150 ÷ 2 = 250mm. 4 × 2 × 250mm = 2000mm is deducted *from the outside girth*, giving 41000mm. Finally, the inner face of the wall has been calculated by taking the outside girth and subtracting 4 × 2 × 325 = 2600mm (the overall width of the wall) to give 40400mm – the same figure as calculated for the inner face in the substructures.[1]

The heights of each element of the outside wall are easily calculated. For the outer skin and cavity, Section A-A shows the ground floor–ceiling height as 2400mm, the floor thickness as 250mm, the height to the window sill as 900mm and the window height as 1200mm (Figure 2.7). Section B-B shows the head of the window to be at the same level as the top of the outer skin, so the height is the sum of these figures at 4750mm. Section A-A shows the inner skin as slightly higher, comprising the ground floor–ceiling of 2400mm, floor construction of 250mm and upper floor–ceiling height of 2400mm. There is a horizontal timber plate on the inner skin, which is shown in the specification to be 50mm deep, so this is subtracted from the height of the blockwork.

AA15/16 shows the items for the outer wall, bringing forward the girths and heights calculated on AA14. First is the external skin of bricks, for which the specification is identical to the facing brickwork in the substructure except that the mortar is described as "**gauged mortar** (1:1:6)". This type of mortar, used above ground level, consists of one part Portland cement, one part

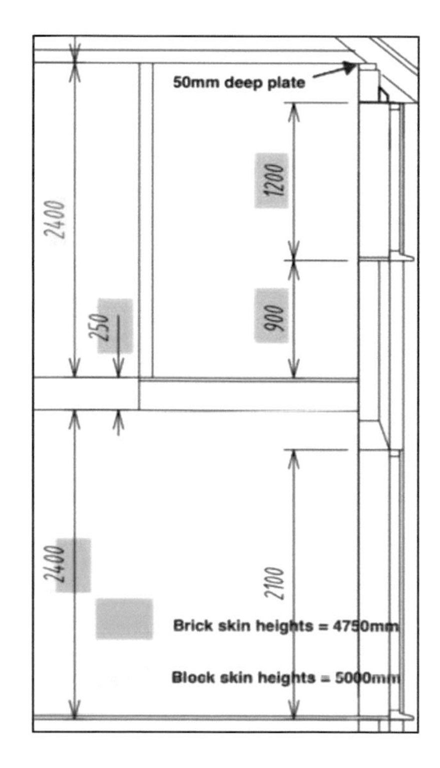

Figure 2.6 Girth of external walling items in superstructure

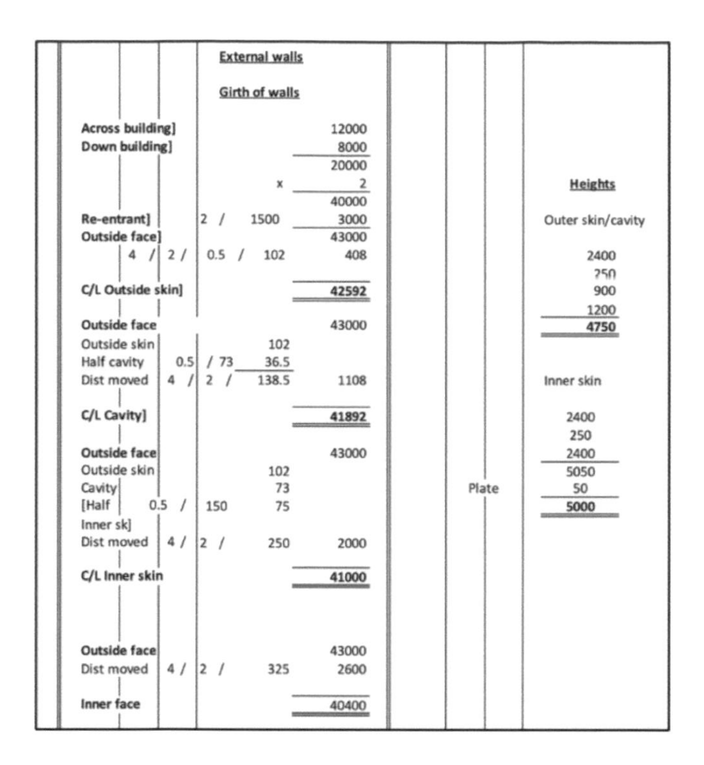

		External walls	
		Girth of walls	
Across building]			12000
Down building]			8000
			20000
		x	2
			40000
Re-entrant]	2 /	1500	3000
Outside face]			43000
	4 / 2 /	0.5 / 102	408
C/L Outside skin]			**42592**
Outside face			43000
Outside skin		102	
Half cavity	0.5 / 73	36.5	
Dist moved	4 / 2 /	138.5	1108
C/L Cavity]			**41892**
Outside face			43000
Outside skin		102	
Cavity		73	
[Half Inner sk]	0.5 /	150	75
Dist moved	4 / 2 /	250	2000
C/L Inner skin			**41000**
Outside face			43000
Dist moved	4 / 2 /	325	2600
Inner face			**40400**

	Heights
	Outer skin/cavity
	2400
	250
	900
	1200
	4750
	Inner skin
	2400
	250
	2400
	5050
Plate	50
	5000

Figure 2.7 Masonry heights

hydrated **lime** (calcium hydroxide) and six parts sand. Compared with cement mortar (one part cement to four parts sand) used in the substructure, gauged mortar is relatively weak. That is. there is less binding cement in proportion to sand. The architect and bricklayer will match the strength of the mortar to that of the bricks, and a 1:1:6 mix suits most. Occasionally, a weaker mix (such as 1:2:9) will be used for very soft bricks. All mixes are adequate for the low loading involved in two-storey domestic construction. As mentioned with the substructure, the higher strength and omission of lime below ground are used to improve acid resistance. Above-ground mixes are gauged (proportioned) to ensure that brick and mortar match, giving uniform strength and moisture absorbency. In historic work, the use of lime predates Portland cement, and a mortar consisting solely of lime and sand is often used. Lime has binding qualities and, depending on its type, will set in the same way as cement (often, however, over a much longer period of time). For modern work, lime is added to mortar to improve its plasticity, making it easier to **work** (use). It will, however, also increase the porosity of the mortar, and this reduces acid resistance, making it unsuitable for use below ground level. As illustrated previously with bricks, bonding and pointing, there is a considerable science and technology behind the specification and use of mortar, only a little of which has been mentioned here.

The item for the cavity is identical to that below ground except that the width is 73mm. "Anded" (&) on to the cavity is the item for insulation. This is specified as glass fibre wall bats treated with resin to provide rigidity and installed by the bricklayers as the work proceeds. Examination of section B-B shows a potential **cold bridge** where the cavity insulation stops, but the inner skin continues up to the plate (Figure 2.8). So, to connect wall and roof insulation and avoid this, a 250-mm-high extra band of insulation is specified and has been measured as an ADD dimension below the main item. Supporting the brief specification for insulation would be more detail on the exact nature and method of fixing. The type of insulation specified is known as "fully filled" and is suitable for walls of moderate exposure. Alternative insulations might involve partially filling the cavity to leave an air space, giving better moisture resistance in exposed situations.

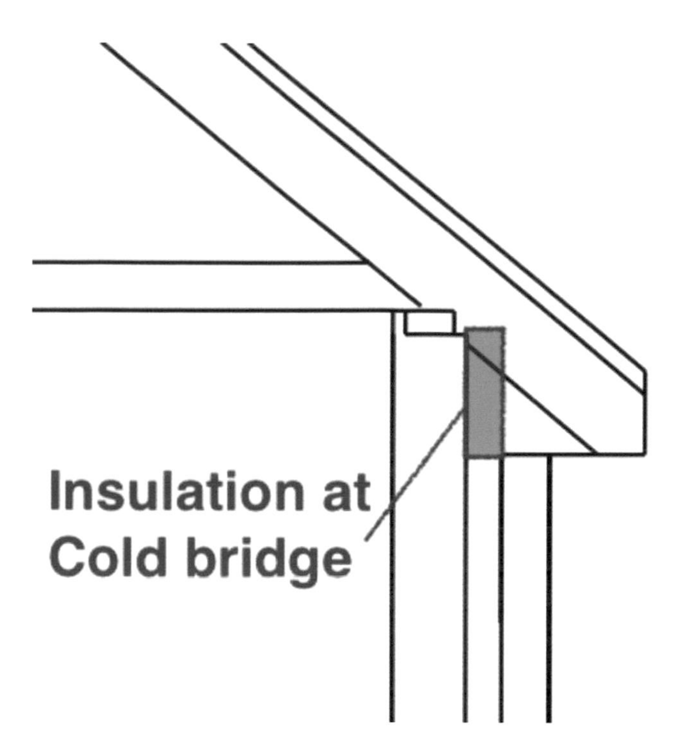

Figure 2.8 Insulation at cold bridge in external wall

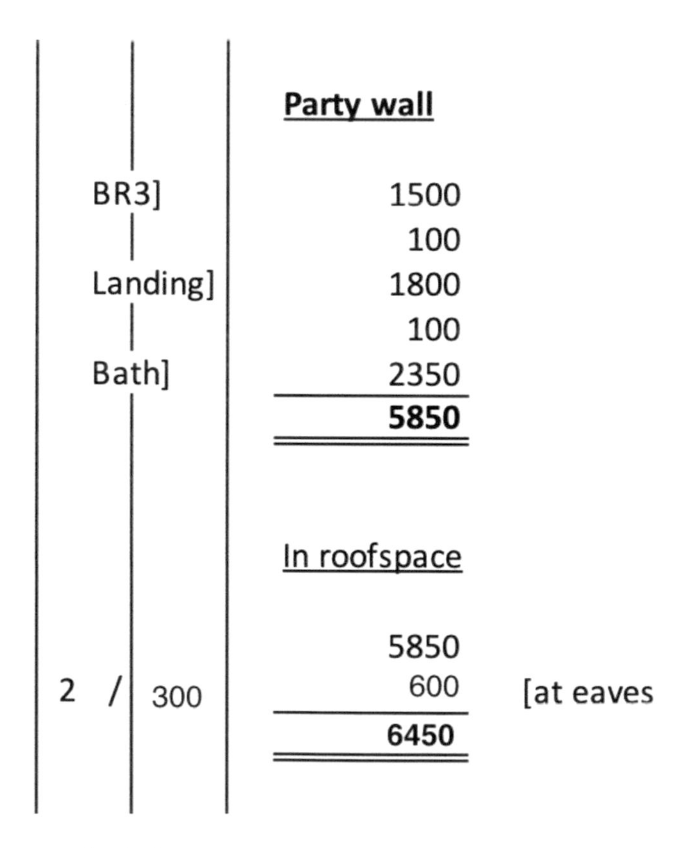

Figure 2.9 Party wall waste calculation

The blockwork inner skin is itemised in the same way as for the substructure but with a varied specification – insulating blocks using gauged mortar. Again, a more detailed specification for the blocks would back up the brief item.

Party wall

AA16 calculates the length of the party wall at 5850mm based on the first-floor dimensions (across bedroom 3, the landing and the bathroom plus the thickness of two partitions). This calculation was previously carried out for the substructure party wall lengths, and re-calculation would not normally be needed. The party wall extends into the loft, giving rise to an extra triangular area ((a) in Figure 2.10), measured by using the formula for calculating the area of a triangle. Part of the triangular area is shown in Figure 2.10. Note that its base is longer than the party wall by 300mm, as each end projects over the outer wall; 600mm is, therefore, added to the base of the triangle, and the calculation for the walling in the loft is 0.50 × 6450 × 2800mm (half the base times the height). The height of 2800mm is shown in Section A-A. The height of the party wall below loft level has been measured as the same as the inner skin of cavity walling, but the timber plate (deducted to give the 5000mm height for the inner skin) does not extend along the party wall, so a small amount of extra party walling ((b) in Figure 2.10 – 6450 × 50mm) has been added. Remember that each skin of party walling is measured separately, so the dimensions are all timesed by 2.

Partitions

Loadbearing partitions are the same height as the inner skin at 5000mm, and their length has previously been calculated at 6450mm. The re-calculation of this length on AA18 is as a check. The relevant lengths are shown on the ground floor as 3350mm (living room/dining room partition – red in Figure 2.11), 2100mm (kitchen/hall partition – blue in Figure 2.11) and 100mm for the thickness of one partition (yellow in Figure 2.11). This does not give the length of the partition where it runs N–S at the living room doorway (green in Figure 2.11), and this has been

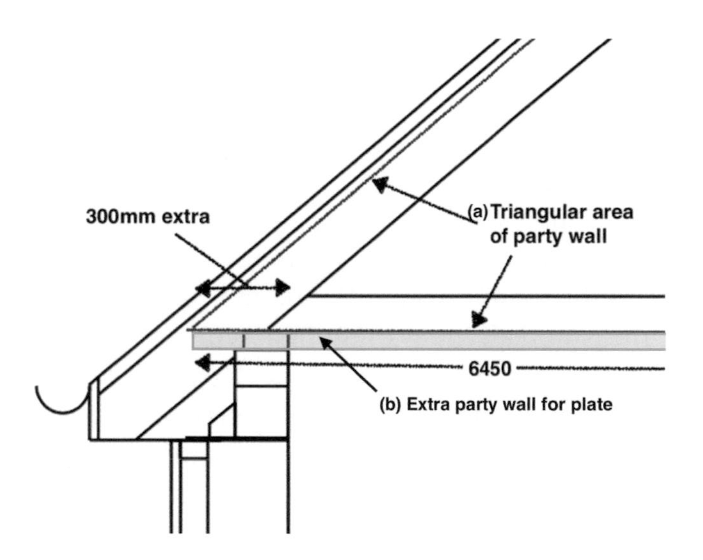

Figure 2.10 Triangular party wall in roof

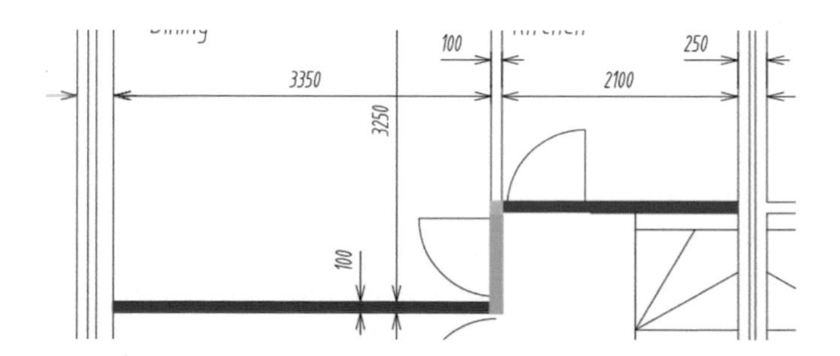

Figure 2.11 Partition lengths

calculated from figures on the first floor (in the substructure, this length was scaled). The distance from inside the external wall at bedroom 1 to the wall behind the staircase is 4900mm, comprising 1500mm for the re-entrant, 1500mm for bedroom 3, 1800mm for the staircase and 100mm for the bedroom 3/staircase partition. As the wall behind the staircase is directly above the wall between the kitchen and hall, then the N–S length of partition at the living room doorway is 4900mm minus the N–S length of the living room of 4000mm, giving 900mm (the same as the scaled length!).

The lengths of the non-loadbearing partitions on ground and first floors have been calculated separately on AA18. The partition between kitchen and dining room is the N–S depth of the dining room of 3250mm minus the 900mm doorway, giving 2350mm. Added to this is the length of the partition between living room and hall, amounting to the N–S depth of the living room of 4000mm minus the 1500mm re-entrant. On the first floor, 2350mm is the length of the bathroom/bedroom 2 partition (the same length as the kitchen/dining room partition), 2100mm is the length of the bedroom 3/landing partition (taken from the ground floor dimensions) and 2500mm is the length of the partition between bedroom 1 and bedroom 3/landing (the same length as the living room/hall partition). The height of non-loadbearing partitions is 2400mm between floor and ceiling with no deduction for a plate. As mentioned earlier, the non-loadbearing partitions are inserted after floors and ceilings are fixed, and plates are not needed. There are two separate items for loadbearing and non-loadbearing partitions on AA18 and AA19, as the specification is different. However, the overall configuration and NRM2 references for all partition items in both substructure and superstructure are similar. Note that all the partitions (unlike external and party walls) are timesed up (multiplied) by 2, as there are two dwellings in the building.

Adjustments for openings

AA20–AA23 deal with adjustments for openings. As mentioned at the start of this section, it is usual to take off walling over all the openings and come back later to make adjustments – deduct areas for windows and doors and add back lintels, arches, sills and jambs. In some approaches to taking off, the adjustments are actually made at the same time as the windows and doors are measured. Thus, taking off windows, doors, frames and thresholds also involves measuring the deduction for walling. More modern sequences of taking off measure the element "net" (that is, the adjustment is made with the main taking off), and that is the approach here. Note, however, that the taker off has still measured all the walling first, with the adjustments afterwards, and not tried to make deductions at the same time as dealing with overall quantities.

The external openings are dealt with first and involve straightforward deductions for the components of the external wall – facing bricks, cavity, cavity insulation and blockwork. It is usual for all the widths of openings to be dimensioned or at least shown on the floor plans but unusual for all the depths of openings to be shown on sections. Either the depth of openings will be determined from window **schedules** (spreadsheets itemising the size and type of all the windows and doors) or, as in our example, determined by scaling from elevations. The taker off can easily count the number of openings for the whole house, and this has been noted as "8 total openings". This ensures that, although some errors are possible in the detailed sizes, at least the overall number of openings is correct. The number of each size of opening has also been identified in the timesing, even if this is of just one unit, and all the openings have been timesed further by 2, as there are two dwellings in the building. This method of entering dimensions minimises errors (by systematically identifying to the taker off that the correct number of units has been adjusted) and makes it easy to follow the thought process and method of the taker off if changes are made in the future. Note also the use of signposts against all dimensions – LR, DR and so on. This allows any interested person following the taking off to identify what has been measured where.

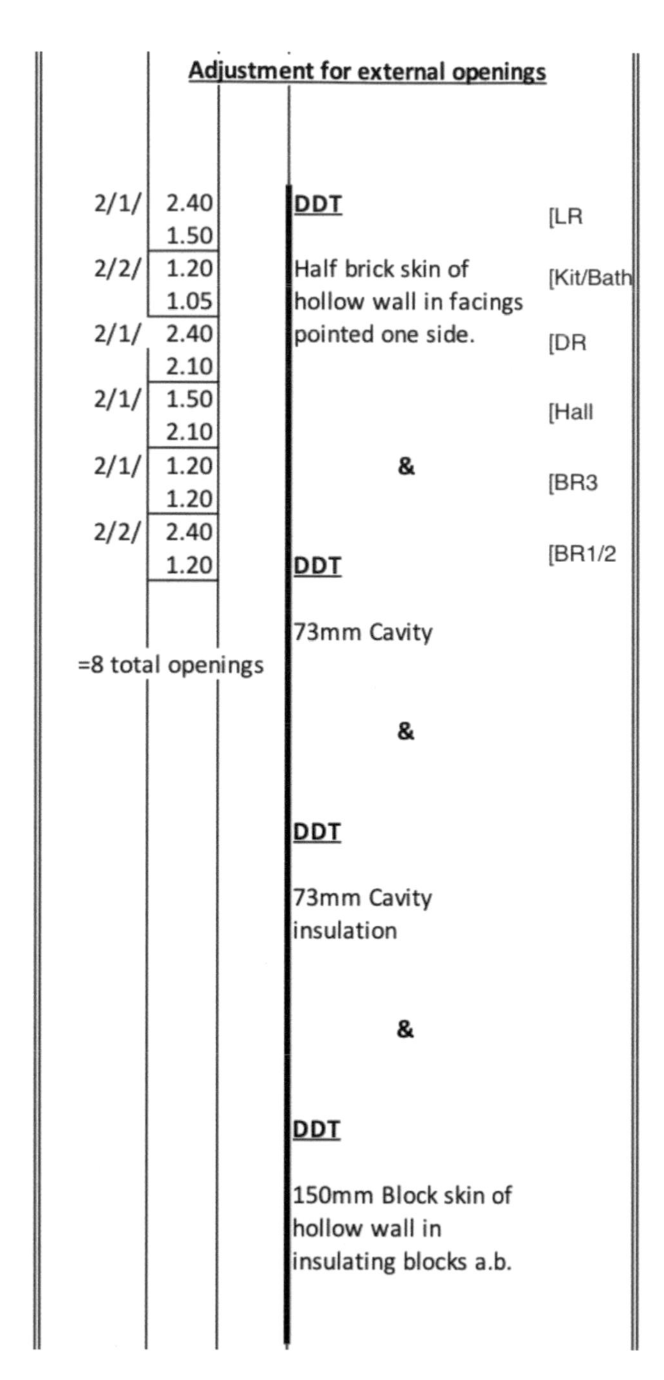

Figure 2.12 External wall adjustment for openings

There are just two opening perimeter items to measure, closing the jambs at the vertical edges and providing lintels over openings. There is no further treatment (such as an arch) over the openings and no brick sill or cavity closer below (the window boards or door units themselves perform these functions for this design). NRM2 14.12.1 requires that the perimeter treatment of closing cavities be measured "extra over" in linear metres, giving a dimensioned description. As cavity closers are standard units, the detailed dimensions, profile and method of fixing will be given in national standards or manufacturer's literature, and reference can be made to these rather than giving extensive information in the items. "Extra over" means that the item is measured without adjustment for any brickwork, blockwork, cavity or cavity insulation displaced. So, if the cavity closer displaced, for example, 50mm of insulation, no deduction would be made to the insulation.

The lintels used for the building are of two types, a cavity wall lintel (Figures 2.1 and 2.13) for ground floor openings and an eaves lintel (Figures 2.2 and 2.14) for first floor openings. They are standard pattern treated steel units made to suit the **span** (width) of the opening and to give a **bearing** over the adjacent walling. Lintels are numbered items and can be specified by length or span (for the latter case, the estimator will add on a bearing at each end to give the length of the lintel). NRM2 14.25.1.1.1 treats steel lintels as "Proprietary and individual spot items", and in this example, they have been specified by reference to a manufacturer's catalogue.

It is usual for the taker off to check the item shown on the drawing and described in the specification against the catalogue reference to ensure that it is accurately described and that every important cost-significant variable is allowed for. Figure 2.13 shows in section the cavity wall lintel, used for ground floor openings. The inner part of the lintel is designed to take masonry, floor and roof loadings, and the outer flange of the lintel (towards the outside face of the wall) is designed to discretely support the outer skin of brickwork. Figure 2.14 shows the eaves lintel. In section it is similar to a cavity wall lintel but is designed to fit directly below the roof construction. Both types of lintels are itemised according to their respective lengths, as shown on AA21 and AA22.

AA23 deals with adjustments for internal partitions. This is very straightforward, but it is worth noting the way deductions have been timesed up. For the loadbearing partitions, the taker off has timesed the opening 800 × 2000mm by "2+2" to indicate two openings on the ground floor and two on the first floor. For the non-loadbearing partitions, the timesing is similar, but "1+2" – one opening on the ground floor and two on the first floor. The timesing figures could have simply been entered as 8 for loadbearing openings and 6 for non-loadbearing openings, but breaking the calculations down allows the figures to be checked and indicates to those following the taking off how the figures have been obtained in relation to the positions of openings in the building. The lintels used over openings in the partitions are of precast prestressed concrete, measured in accordance with the Precast Concrete section of NRM2–13.1.1.1. They have been measured stating the length, calculated by allowing a bearing of 100mm beyond the opening at each end. No

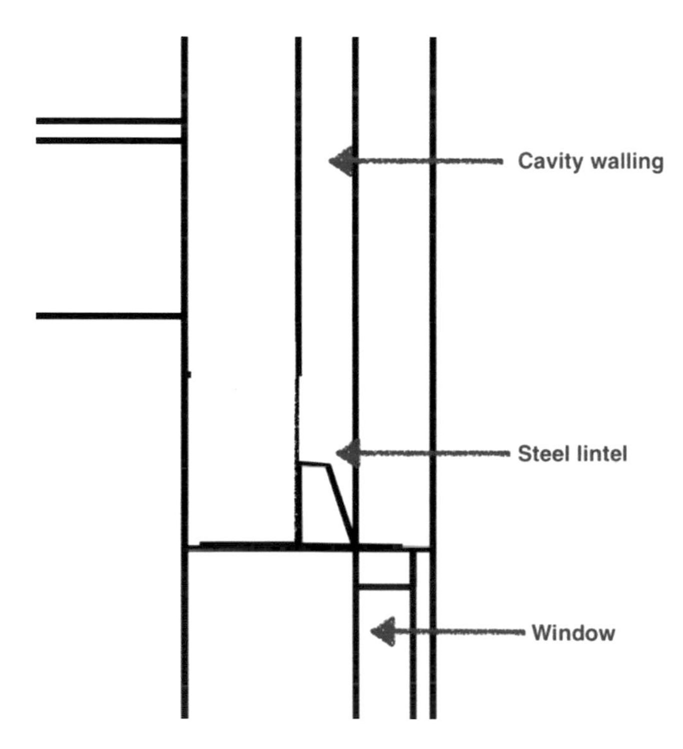

Figure 2.13 Lintel in cavity walling

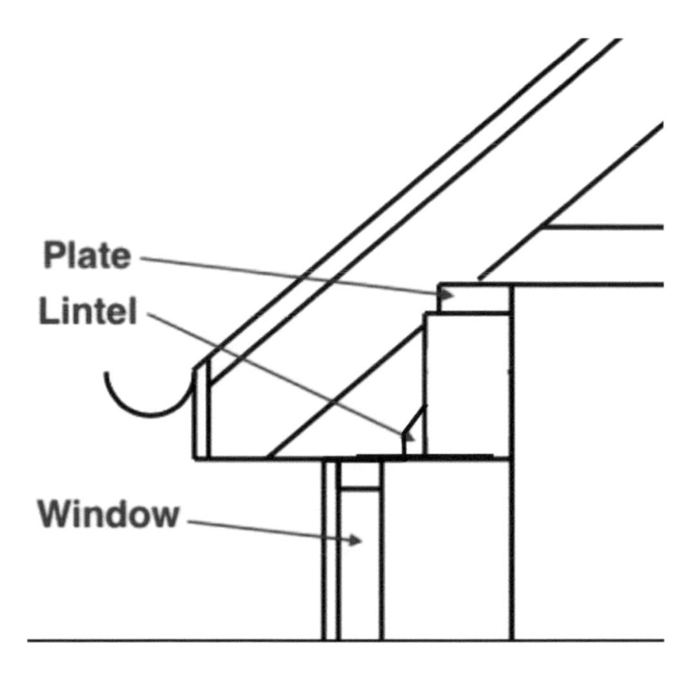

Figure 2.14 Lintel at eaves

deduction has been made for the blocks displaced, as the displacement area for each lintel is less than 0.50m^2 (General rules to NRM2 Section 14, notes on page 198).

Key points covered in this chapter:

- Outline of the technology of domestic external and internal walling
- Sequence of construction for loadbearing and non-loadbearing walls
- Calculating centre-line girths for the elements of masonry cavity walling
- Identifying items for taking off external and internal walling
- Constructing item descriptions for walling using NRM2
- Measuring openings – the form for deducting quantities
- Perimeter items – lintels, jambs and sills
- Setting out taking off for walling and partitions using accepted conventions
- Applying explanations to the example

Note

1 NB: where brick facings have been used in the **substructures**, the girths are the same as for the superstructure – that is, the substructure facing brick girth should be 42592mm, and the cavity should be 41892mm (not as in the substructure taking off at 42600mm and 41900mm, respectively). This amounts to a very small over-measurement, which the taker off has ignored for simplicity and because the overall quantity when rounded to whole units (m^2) will not be affected.

3 Roof construction

AA24–AA28 shows the taking off for roof construction. Take a look at the roof construction plan Q1/A4/2667/5c – at first sight it appears extremely complex. However, there is a considerable amount of regularity, repetition and symmetry in the design of the roof, and it is not nearly as complex as it appears. Before tackling the taking off, however, it would be useful to review the design and technology of domestic roof construction. Modern roofs are usually fabricated using factory-made timber **trussed rafters**. These consist of rigid triangular composite units, designed to span from one side of a typical house to the other, without gaining support from any beams or partitions within the upper floor area. The material used for the trusses is of good-quality softwood, but the main structural distinguishing feature is that all the joints between softwood members making up the truss are rigid, so the truss acts as a single unit. It is this unitary nature of trusses that allow them to span considerable distances and allows quick and easy construction.

Figures 3.1 and 3.2 illustrate standard truss designs as used by volume house builders. Modern trusses are designed, priced and supplied by the unit based on outline designs provided by an architect. Taking off is, therefore, very straightforward compared with the more traditional design used in our project. The taker off simply has to identify and count each type of truss in accordance with the requirements of NRM2, 16.2.1.1.1, based on its overall dimensions.

Trussed rafter construction has its limitations and is less able to accommodate intricate roof shapes, such as **hipped roofs** and **dormers**. Although these features can be engineered using trusses, low levels of repetition reduce their economic advantage. For historic work, trussed rafter construction may not be appropriate, and for small works, alterations and extensions, the flexibility of traditional construction may outweigh any marginal economic benefit of trusses.

The roof in our project is based on traditional construction, where each individual timber performs a separate structural function, and the roof is built from timbers cut and fixed on site. A **cut** or **built-up** roof are common names for this type of construction. As each timber acts separately rather than as a unified truss, a built-up roof is structurally less efficient than trusses. Common timbers, such as rafters and ceiling joists, are able to span about 3–4m, and the roof timbers, therefore, may need intermediate support in the form of loadbearing partitions, struts and purlins. Figure 3.3 shows the members of our traditional built-up roof. As the method of construction was prevalent until the 1950s, this construction can be seen in any older house built before World War 2, particularly the archetypal pre-war semi-detached house.

The hips on our roof are the ends of both the main and small roofs (over bedroom 1), which slope inwards at the same **pitch** (angle) as the front and rear slopes. Figure 3.4 shows the hip generally, together with the timbers used to form it. Figure 3.5 is an internal view of a similar roof. The hip rafter performs a similar function to the main ridge board but slopes downwards from the end of the ridge towards an outer corner. The jack rafters perform a similar function to the (common) rafters but span between the hip rafter and edge of the roof and are of diminishing lengths towards the outer end of the hip rafter.

DOI: 12.01/9781003253129-4

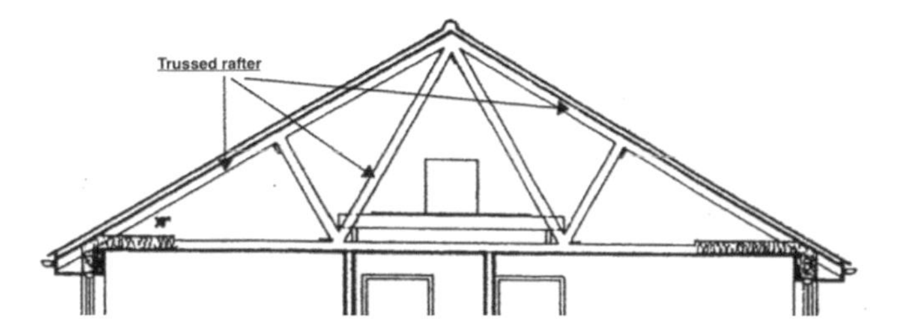

Figure 3.1 Trussed rafter section

Figure 3.2 Trussed rafters in loft

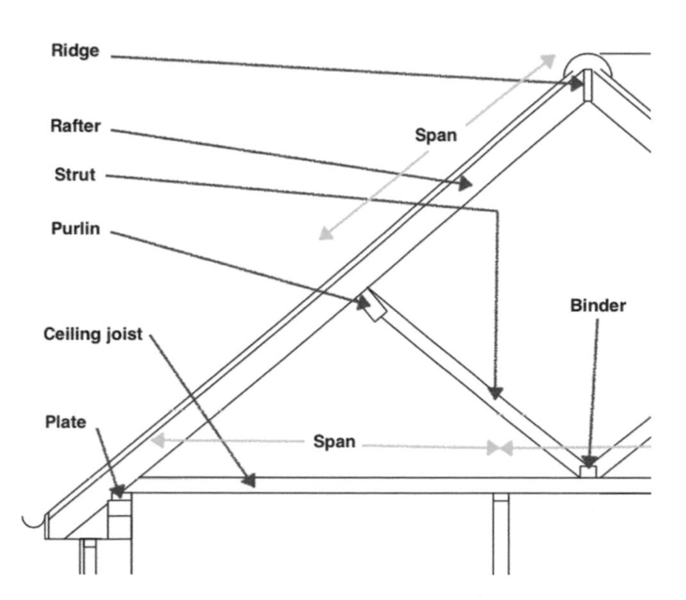

Figure 3.3 Traditional built-up roof

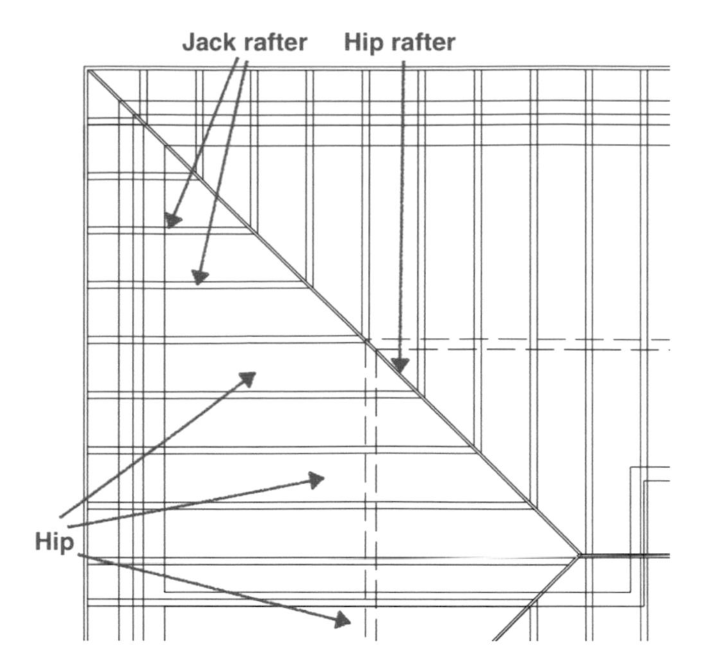

Figure 3.4 Hipped roof construction

Figure 3.5 Interior of hipped roof

As with other aspects of construction covered in this book, there is a depth of science and technology behind this simple explanation, and reference should be made to construction technology texts for more detail.

AA24 of the taking off starts in the usual way by itemising the drawings used and constructing a taking-off list. As we are dealing with lengths of timber, nearly all items in structural carpentry are measured in linear metres. The items involved and brief descriptions are as shown in Table 3.1:

On the right column of AA24, the lengths of the wall plates are calculated. They run all round the top of the inner skin of the outside walls and on top of the loadbearing partitions. The outer dimensions of the whole building are girthed up, and extra is added for the re-entrant at the front. This is the same girthing calculation as carried out in the substructure and walling. The resulting girth is then adjusted to calculate the length of the inner face by deducting 4 × 2 × 325 (the width of the wall), giving 40400mm. This figure has also previously been calculated, both in the

Table 3.1 Taking-off list for roof construction

Plates	Horizontal timber at the top of loadbearing walls to spread load and give even level bedding for rafters and ceiling joists. Plates involve two trades in cutting the timber (by a carpenter/joiner) and bedding the plate (by a mason/bricklayer).
Rafters	Main supports for roof coverings spanning from ridge to plates. Note there is no mention of "jack rafters" – see later for why they are not specifically measured.
Purlins	Intermediate supports for rafters at mid span, themselves spanning between struts Purlins are not needed for the small roofs over bedroom 1, as the rafters span a much shorter distance.
Binders	Tension members to stop hips from spreading outwards, fixed to the ceiling joists and the end of a rafter at the hip.
Strut	Intermediate compression support for purlins transferring loads from the roof to the loadbearing partitions.
Ceiling joist	Main support for ceiling and items in the loft. Ceiling joists also provide tension restraint for the rafters, which otherwise would tend to spread outwards.
Ridge (board)	Separates rafters and provides support for ridge tiles at the apex of the roof.
Hip (rafter)	Similar function to ridge board at sloping hip.
Valley (rafter)	Not shown – similar function to hip rafter but inverted at valley between the small roof inner slopes and the main roof slope.

substructure and external walling, and it is not really necessary to re-calculate it here. The *outside* girth of the plate, not the centre line, is then calculated by adding 4 × 2 × 100mm (the width of the plate). The outside girth is conventionally used in taking off lengths of timber, as they are always lapped or mitred on corners (i.e., timber is always cut to the external length). The length of the plate to the loadbearing partition has previously been calculated as 6450mm, and this is marked has having been brought forward from page AA18.

Timber is normally supplied in lengths of about 3.6–4.5m, meaning it either needs to be jointed or supplied as a special item (at higher cost) if used in longer lengths. If work can be jointed, as it can for the plates, the taker off will not allow extra length for forming the joint (for example, by using a **halved joint**) – an allowance will be made by the estimator. If work cannot be jointed (and there are obviously limits on how long timber can be based on the height of trees), then the cost may be higher. So, if a continuous length of more than 6m is required, it is specified in the taking off. For a rough guide on where timbers are likely to be required in one length, timbers acting as beams – that is spanning between two points – will need to be in a single length. If over 6m long, NRM2 requires that this be stated in the item description (16.1.1.*.9).

The plate is itemised on AA25 with lengths for the outside wall and partition. Note that the plate to the partition is timesed (multiplied) by two, as there are two dwellings. NRM2 covers plates in section 16, Carpentry, subsection 1, primary and structural timbers. Level 1 requires that the **nominal** size of the member be stated. Nominal size means the size of the timber as converted from the log – that is, with an unfinished (often described as a **sawn**) face. In contrast, **finished** size is the size of the timber when planed. This size will be less that the nominal size to the extent of the **planing margin** – the amount that is taken off by planing. Margins are usually around 3mm and are important when finished timber is specified by nominal sizes. Thus a 50 × 100mm nominal planed timber will only actually be a finished size of 44 × 94mm.

Constructional timbers in section 16.1 of NRM2 are assumed to be sawn and of nominal sizes (NRM2, 16 general rules), so the nominal size is the same as the actual. General joinery items (covered in section 22 of NRM2) and floor boarding (covered in 16.4 of NRM2) are normally planed (or **wrought**, to use the traditional term). When specified by nominal sizes, the wrought size will be smaller than the nominal size. Confusingly, architects may specify sizes as nominal or finished, so it is important in taking off and pricing to be aware of this distinction and to accurately describe the item. This will include where nominal sizes are quoted but planed timber required, specifying the planing margin.

Wall plates, as a level 2 descriptor of NRM2, are a separate category, and at level 3, a number of variations on types and treatment of the timber are drawn out. All constructional timber in this building is treated at the yard with preservative, and this is identified in the item description with a further reference to a general specification.

The next item measured is the rafters, and AA25 first considers their length. The main "common" rafters as shown on Section A-A could be scaled, but it is preferable to use figured dimensions and calculation if possible. In this instance, the taker off has used trigonometry to calculate the length using the ceiling joist as base and the **pitch** (angle) of the roof. The calculation for the base is shown on AA25 and involves taking the N–S depth of the building, deducting the re-entrant at the front and dividing this by two. The overhang for the eaves is then added to give the length from the middle of the main roof to end of the rafter (i.e., the **plan length** of the rafter).

The rafter forms a hypotenuse to a right-angled triangle with the ceiling joists as the base and the vertical height of the roof to the ridge (apex) as the opposite side. It is therefore possible to calculate the hypotenuse by using the cosine of the pitch of the roof. The calculation is shown in the right column of AA25, and the taker off has also provided a diagram showing exactly how the calculation has been executed (see Figure 3.6). The resulting figure of 4504mm can be checked for approximate accuracy by scaling the rafter from Section A-A.

The next waste calculation involves calculating the number of *pairs* of rafters to the main roof. The constructional plan of the roof (Q1/A4/2667/5c) shows all the rafters, so it is possible to count these. However, in this instance, the taker off has calculated the number based on the overall length of the roof and the spacing of the rafters. The building is 12000mm long, and, allowing for a 200mm overhang for the eaves at each end, the roof is 12400mm long. As the rafters are specified to be spaced at 400mm **centres** (the distance from centre to centre of each rafter), the number of rafters is 12400/400 = 31 exactly. With a hipped roof, there is no rafter at the end of the roof, so one must be deducted to get the exact number of 30. In this example, the number can be confirmed to be correct by actually counting from the constructional plan. Note that although the roof has hipped ends involving jack rafters of varying lengths (very short at the outer ends and almost full length at the inner ends of the hips), the taker off has ignored this. This is because,

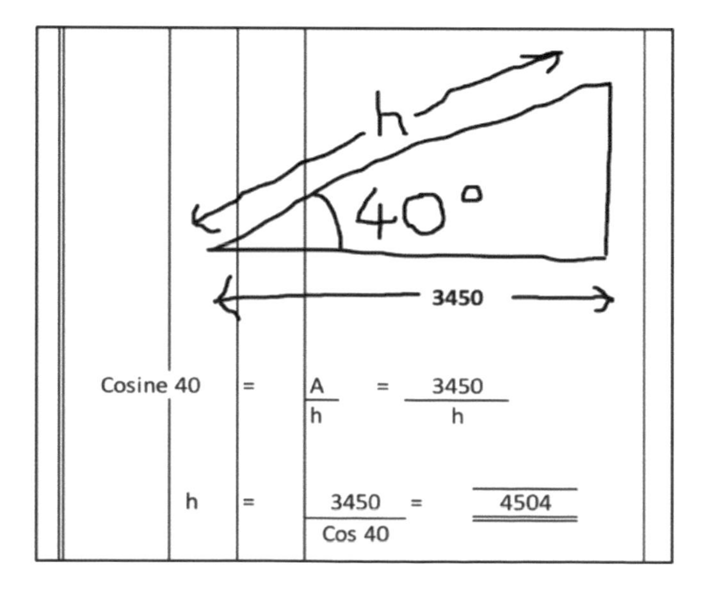

Figure 3.6 Length of rafter by trigonometry

although there is more cutting, the *length* of timber involved is exactly the same for a hipped as a **gabled** roof. Another way of describing this is that the length of rafter per m^2 of roof area is the same whether measured over the main roof area or the hipped area (2.5m per m^2 of roof area for rafters at 400mm centres).

The length and number of pairs of rafters to the small roofs over bedroom 1 are calculated in exactly the same way as for the main roof. The length of these roofs is 1500mm, and when divided by the 400-mm centres of the rafters, there are 3.75 pairs. It is not possible to install part of a rafter, so the carpenter will reduce the centres and round the number of rafters up to the next whole number – in this case 4. As there is only one hipped end for these roofs, the number of rafters is not reduced for the hipped end. Examination of the construction plan confirms that the correct number of pairs of rafters is 4. Note that there are no extra rafters for the parts of these roofs that cut (extend) back into the main roof, as rafters have already been measured here with the main roof.

AA26 shows the item for the rafters, and signposts indicate that the first dimension is the main rafters, then the rafters to the small projecting roofs over bedroom 1. The following two dimensions are an allowance made for an extra rafter to each hip directly opposite the ridge. This is a conventional allowance, not necessarily shown on drawings, to help in setting out a hipped roof (i.e., the carpenter will first set out the framework of ridge, hip rafters and one common rafter before filling in the rest of the rafters).

The description for the item is drawn from NRM2, 16.1.1.1.8 – Carpentry, primary or structural timbers, nominal size, a category of rafters and associated roof timbers and the off-site preservative treatment. Rafters, struts, ridge, hip and valley rafters are all classed as "rafters and associated roof timbers" in this example. The **cross-section** size in taking off and for construction is normally given as breadth × depth of member – unlike structural engineering, where the engineer will often specify by depth × breadth (the depth of a member usually being the key dimension in resisting loading).

The next item is the purlins. These provide support to the rafters at mid-span. They run round the whole of the main roof but are not needed for the smaller roofs projecting over bedroom 1. Their length is easily calculated, as, although there are four lengths of purlin at mid span of the main roof slopes (front, rear and two hips), their equivalent length is the same as the overall internal length of the roof × 2. This is illustrated in Figure 3.7. The lengths taken off are the blue lines. As the roof is symmetrical, with the purlins at mid span of the rafters, this length is the equivalent to the line of the purlins shown in red.

The binders run along the main and small roofs, join to the foot of a rafter and are nailed to the ceiling joists. As mentioned, their function with this type of roof is to provide restraint to the hip. This is necessary, as the ceiling joists, which provide restraint to the rafters on the main slopes, run parallel to the hips. Without restraining ceiling joist or binders, the weight of tiles, rafters and **live loads** would force the hip outwards at the eaves. The binder is shown as being the same length as the overall length of the building, 12000mm, and the nearest NRM2 classification is as a beam at 16.1.1.5.8. It can be jointed, as it is only acting as a tension member, so it is not necessary to describe it as in one length. The binders to the small roofs over bedroom 1 are a different cross section and length but otherwise the same category in NRM2.

Struts support the purlins at mid-span, which in turn are at mid span to the rafters. Intuitively, therefore, their length will be the same as half the span of a rafter. Expressed mathematically, the bottom span of the rafter from the purlin downwards, the strut and ceiling joist form an isosceles triangle, of which the strut is the same length as the half rafter. The specification states that there is one strut to 4 rafters, and there are 30 rafters. This gives 7 struts with a remainder, so their number is rounded up to 8. There is no separate NRM2 category for struts, so they are classed as "rafters and associated roof timbers".

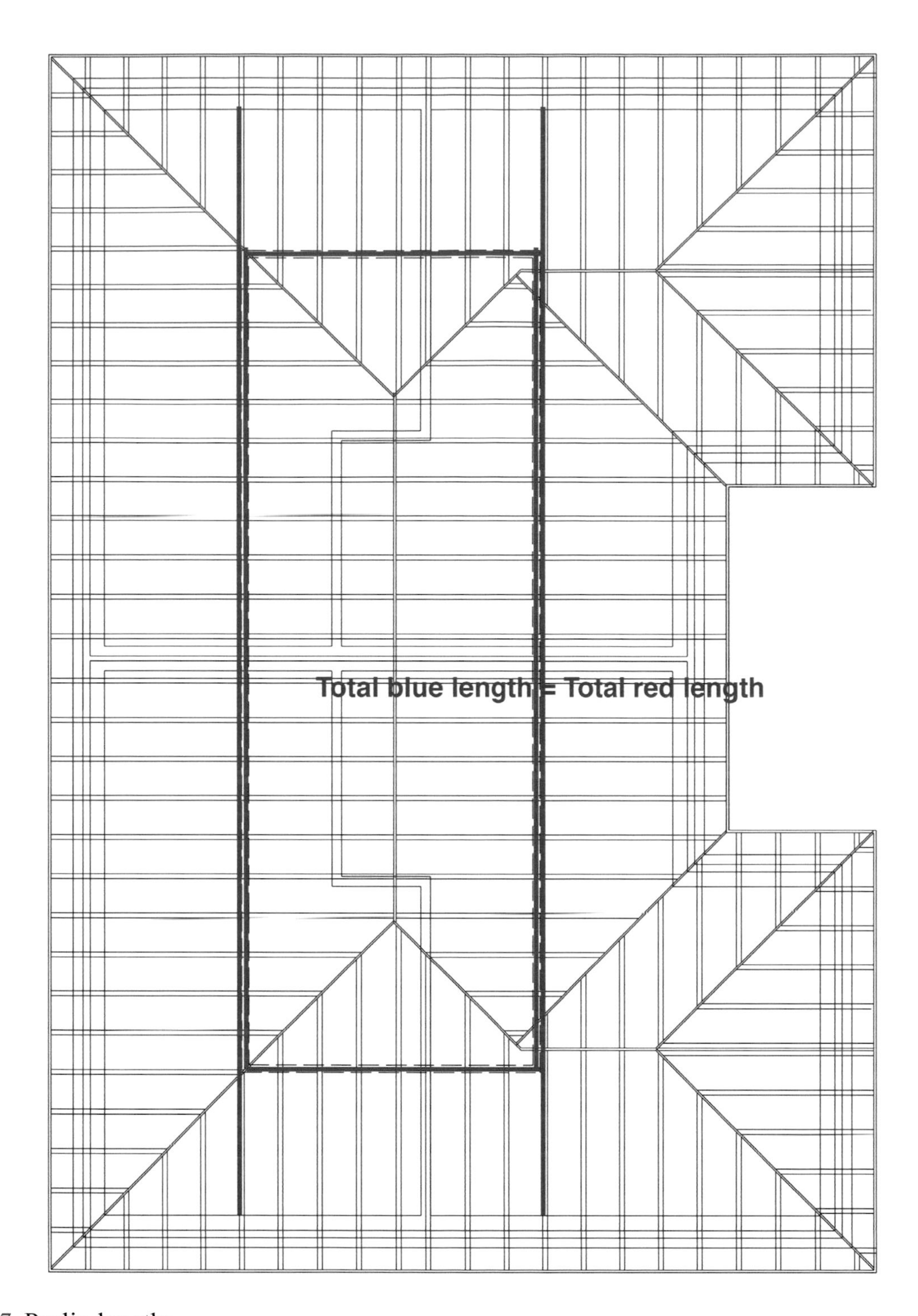

Figure 3.7 Purlin lengths

The ceiling joists are the same length as the N–S depth of the building at the outside face of the wall, previously calculated at 6500mm for the main roof and 4000mm for the small roofs over bedroom 1. There are the same number of joists as pairs of rafters, as each ceiling joist is tied to a rafter at the foot. They are classed as "roof and floor joists" under NRM2 16.1.1.4.8. As with the binders, there is no necessity for ceiling joists to be in one length (they are often jointed over a loadbearing partition).

AA28 shows the taking off for ridge, hip and valley rafters. First, the ridge can easily be calculated, as its length is the length of the roof, minus the length of a rafter on plan at each end. This is illustrated in Figure 3.8. The plan length of the rafter has previously been calculated at 3450mm, and this is deducted twice, as there are hips at each end of the roof.

For the small roofs, ridge, roof edge, hip rafter and valley rafter form a parallelogram where opposite sides are equal. So, the length of the ridges to the small roofs is the same as the lengths of the edges, which is the same as the re-entrant at 1500mm.

The lengths of hip rafters are not shown on any plan or section, so they must be calculated. This can be done by using Pythagoras. The rafter length is known (4504mm), as is the plan length of the rafter (3450mm). Forming a right-angled triangle of rafter length on plan, rafter length on slope and hip rafter, the hip rafter length can be calculated as the root of the sum of the square of the two other lengths. This is illustrated in Figure 3.9.

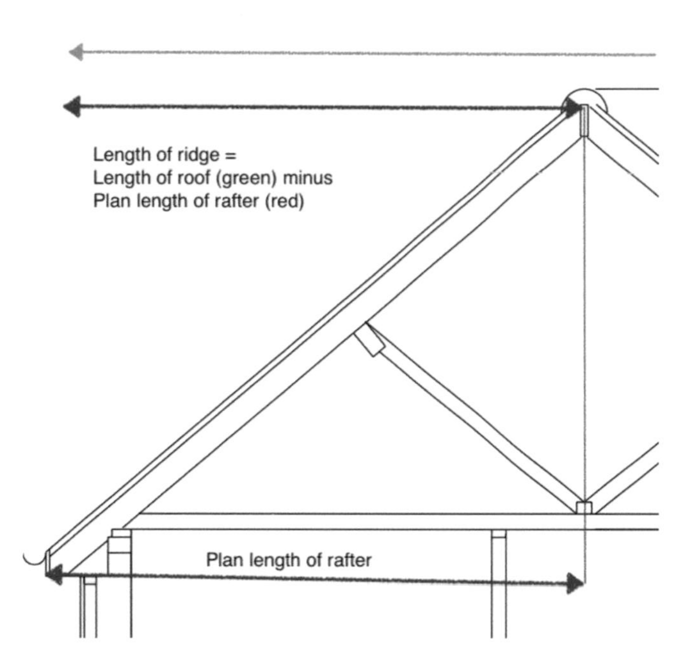

Figure 3.8 Length of ridge

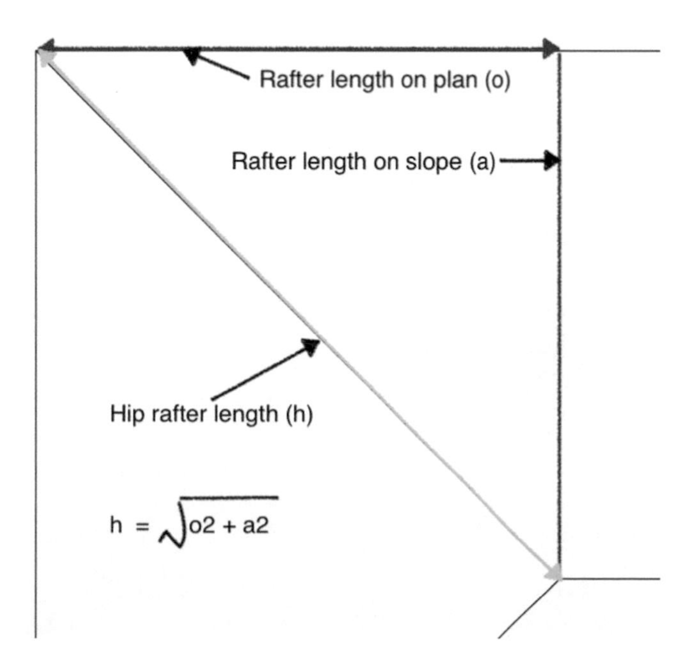

Figure 3.9 Length of hip rafter

The small hip rafters to the roofs over bedroom 1 can be calculated in the same way, but for these hip rafters, only two are taken. This is because the outer hip rafters have already been measured with the hips to the main roof, as illustrated in Figure 3.10.

The final item is for the valley rafters at the inner edge of the small roofs over bedroom 1 (Figure 3.11). These timbers, although the same length as the small hip rafters, act as beams, supporting the jack rafters cut into each side. They are, therefore, of more substantial timber.

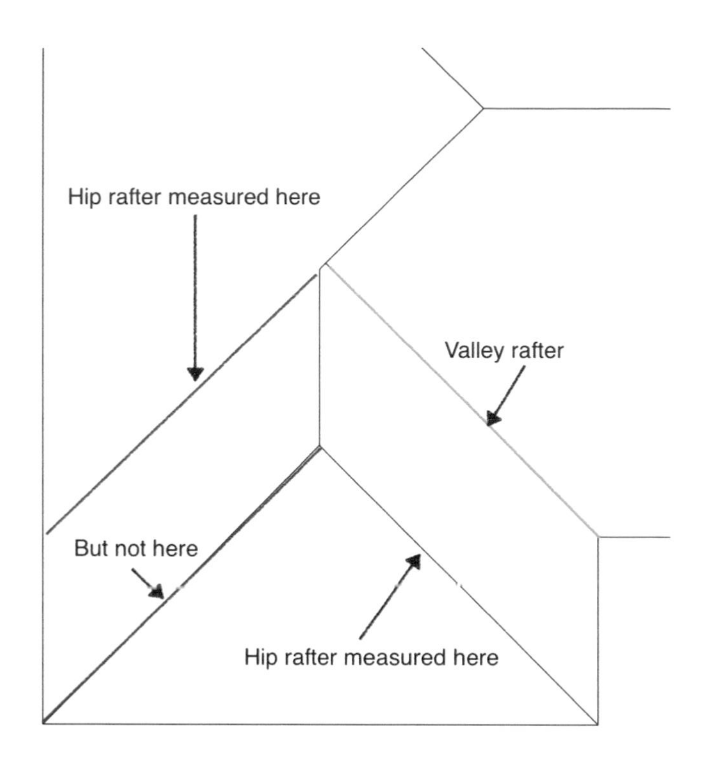

Figure 3.10 Hip rafters to small roofs

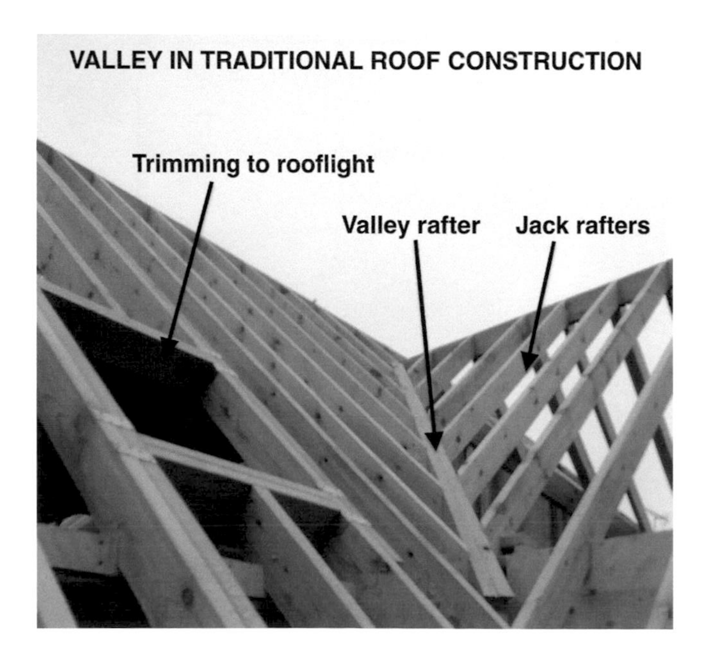

Figure 3.11 Valley rafter

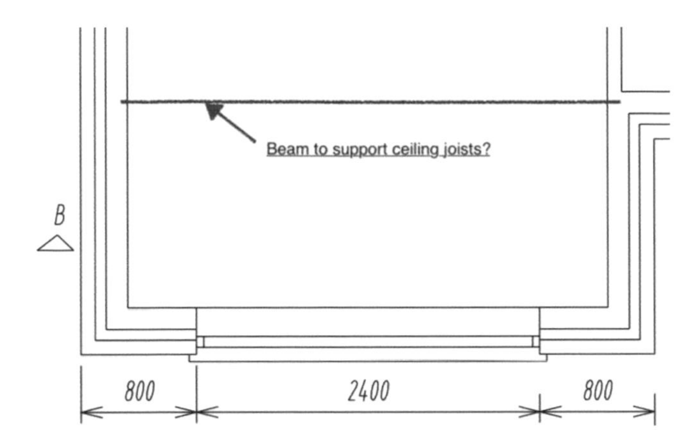

Figure 3.12 Beam to support ends of ceiling joists

Experienced takers off have a very good knowledge of construction technology, developed by continually measuring construction details. The taker off has, at the end of the roof construction section, noted an item to measure of a "*beam to support ceiling joists at bedroom 1 projection*". Based on knowledge of technology, the taker off has noted that the main ceiling joists are not supported at the ends where bedroom 1 projects forwards. That is, the detail shown at the right-hand eaves in Section A-A does not apply to a cross-section taken in the same direction through bedroom 1. There is no wall or plate to support the ceiling joists over the projection to the bedroom.

At the same time as the note is added, the taker off will put a query to the architect concerning the omission and seek clarification on any item to be included. In this way, the taker off is acting to ensure that the design and measurement are complete, fully reflecting the work involved and the cost of the project.

Key points covered in this chapter:

- Outline of the technology of modern and traditional domestic roof construction
- Names of traditional roof timbers
- Use of basic geometry to measure the lengths of timbers, including trigonometry, Pythagoras, isosceles triangles and parallelograms
- Calculating outside length girths for timber plates
- Identifying items for taking off roof construction
- Constructing item descriptions for roof construction using NRM2
- Item to Measure notes and their use in taking off
- Applying explanations to the example

4 Roof coverings

AA29–AA32 show the taking off for roof coverings. Notice that there are very few waste calculations in this section and no preparatory items, such as a drawing or taking-off list. This is because most calculations and preparatory items are common to the roof construction and do not need repeating in this section. Although the roof construction is taken off before the coverings in these examples, many takers off prefer to measure the coverings first. This gives a good introductory overview of the construction, allowing the measurement of the latter to be rationalised. Roof coverings are a section of taking off, very like other surface treatments (such as wall and floor finishings, ceilings and paving), consisting primarily of areas and lengths. The main items are areas measured in m^2, with the treatment of boundaries measured in m and the occasional enumerated item at intersections.

Before considering the taking off in detail it is worth reviewing the technology of roof coverings as used in housing. Thatch is the original vernacular covering for roofs in the British Isles, supplemented by slate and stone. Although the Greeks and Romans had fired clay roof tiles, fired clay was not introduced in Britain until about the 14th century. Early tiles were fired flat pieces of clay of varying size and thickness, overlapped sufficiently to form a continuous water-shedding covering. Having varying sizes proved most inconvenient, and led to one of the earliest pieces of surviving standardisation legislation in 1477.[1] Plain clay and concrete tiles have been standardised since that date to what is now 265 × 165 × 15mm. They are called "plain", as they are nominally flat and allow water to run off the edges. This means that they must be lapped sufficiently to stop this happening and water entering the building – vertical joints are staggered like stretcher bond brickwork, and there must be at least two thicknesses of tile over all the roof area, with a bit extra on the ends (the **lap**, giving three thicknesses) to allow a margin for windy weather. Figure 4.1 shows **plain tiling**, a **bonnet hip**, a **verge** and a tiled **valley** (see subsequently), and Figure 4.2 shows a section through a plain tiled roof for an historic building. Modern plain tiles are not quite flat but cambered in two directions to reduce the likelihood of seepage by capillary action.

Greek and Roman tiles are not flat but consist of two types – *tegulae:* trough tiles with turned-up edges, and *imbrices:* capping tiles fixed over the joints between the tegulae. The combination of trough and capping means that water cannot run off the edges of the tiles and is directed down the slope. This type of tiling is the forerunner of modern interlocking tiles (illustrated in Figures 4.3 and 4.4) used widely by volume house builders in the United Kingdom. Stopping water running off the edges of the tiles means that they only need to be covered at the edges (by imbrices or with modern tiles, the interlocking edges shown in Figure 4.4) and are known as **single lap** tiles as opposed to **double lap** plain tiles. Having single lap tiles makes the covering lighter, and the interlocking edges mean that the roof can be pitched to a shallower angle and remain water resistant.

In current construction, interlocking concrete or clay tiling predominates, but plain tiles are still common in high-quality housing, notable buildings and historic repair and restoration. The

DOI: 12.01/9781003253129-5

Figure 4.1 Plain tile roof coverings

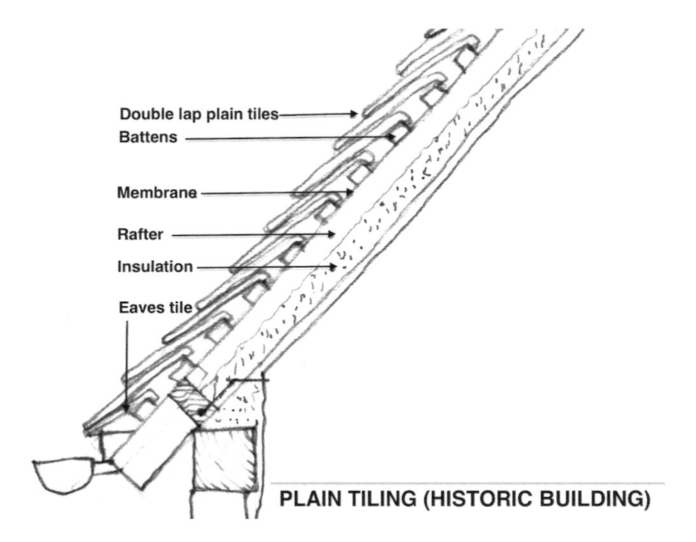

Figure 4.2 Section through plain tiling

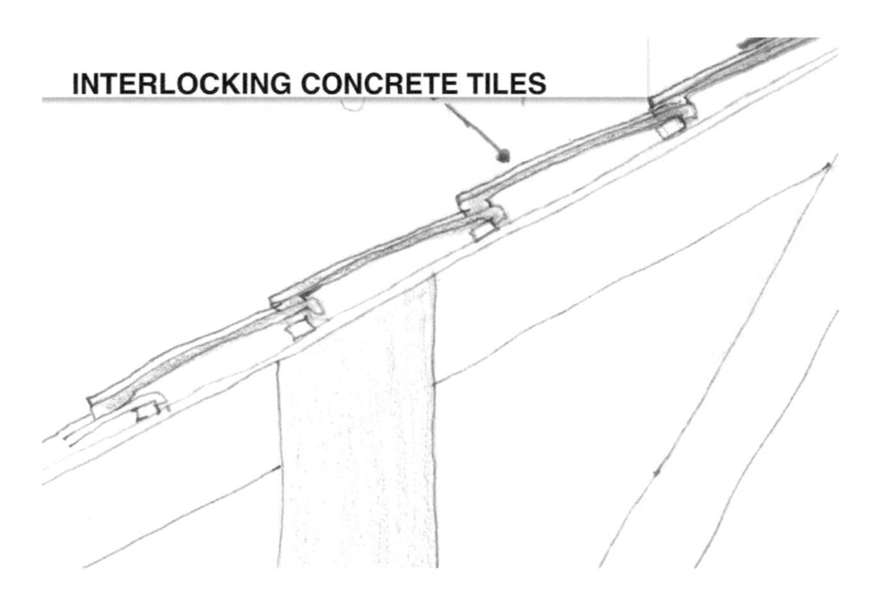

Figure 4.3 Section through interlocking tiling

technology of tiles, as manufactured items, is supported by accessible technical detail, often produced by the tiling manufacturers but supplemented by national and international standards. Specifying, therefore, usually refers to this reference material.

There is also an established slating and stone roofing industry. Slates were widely used for volume housing until the 20th century and are still used for new work, repairs, extensions and historic restoration. Slates and stones are flat and have similar characteristics to plain tiles – they must be double lapped. Sizes are not governed by statute, and there are many variations. For some types (for example, Welsh slates), there is a limited range of recognised common sizes. Stones, such as Cotswold stone, are usually graded by size, and many roofs are built with varying-sized stones.[2]

All types of tiles, slates or stones are set on battens (narrow strips of timber), usually by nailing. Some early types of tiles are hooked over the battens using timber pegs (oak or similar, as illustrated in Figure 4.5), and modern plain and interlocking tiles have nibs cast into the undersurface allowing them to be hooked over the battens (Figure 4.6). For these tiles, there may be

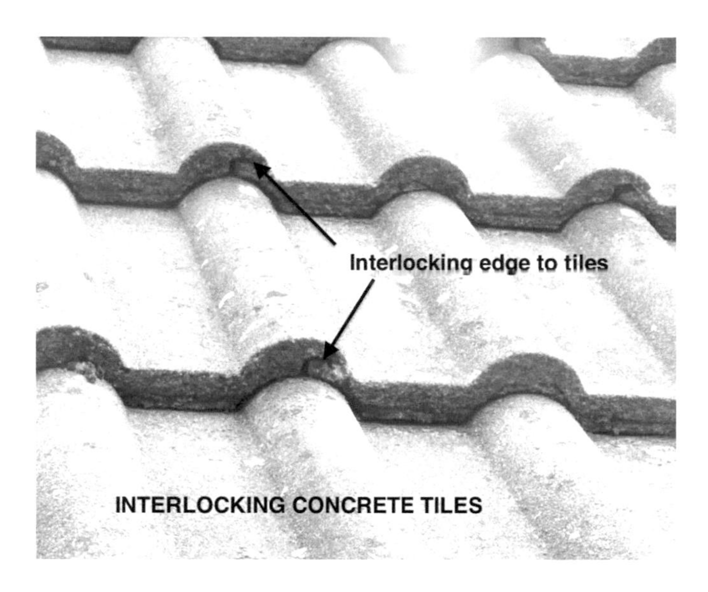

Figure 4.4 Interlocking tiling

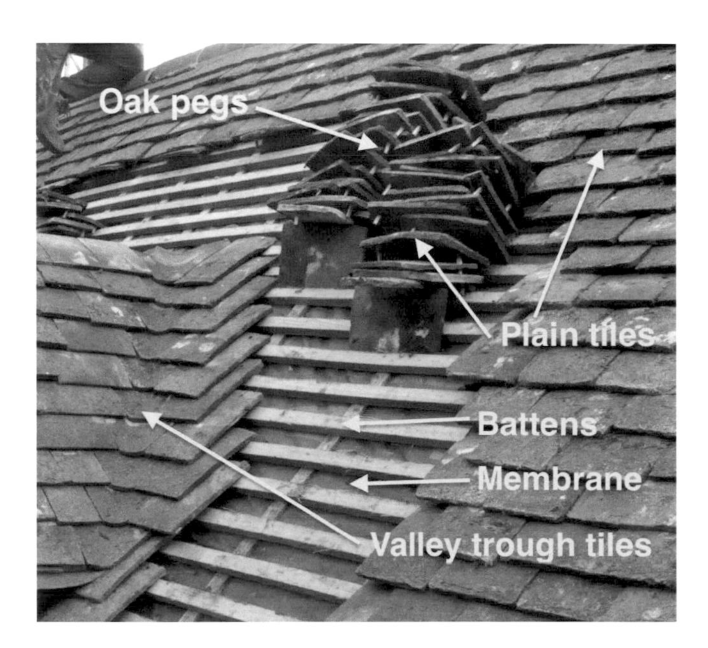

Figure 4.5 Pegged plain tiling details

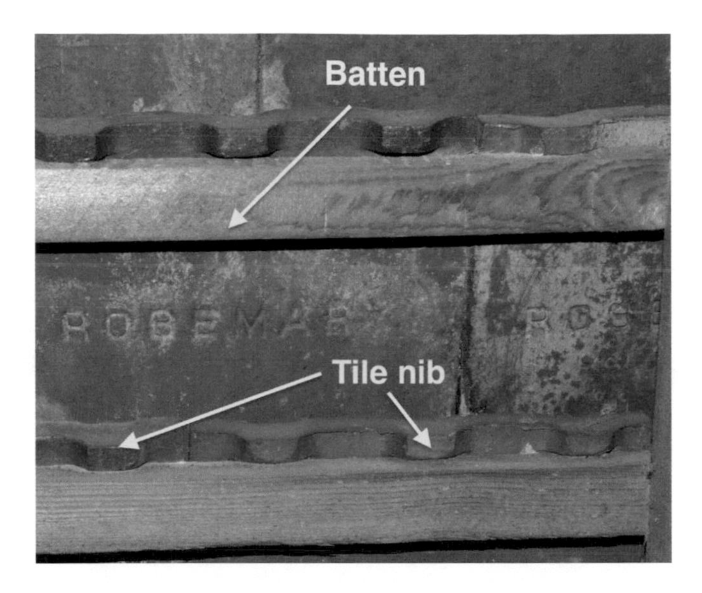

Figure 4.6 Nibs to underside of plain tiles

supplementary nail holes allowing full or intermittent nailing depending on their degree of exposure to the wind, or the tiles may just be held in place by gravity. For our example, tiles are nailed to battens every fourth course, giving a good compromise between economy and performance. In current construction, there is a waterproof underlay membrane between the battens and rafters. This acts as a secondary form of waterproofing, particularly against driven rain and snow. It also improves thermal insulation and keeps the loft area clean. Modern **under-tiling** or **under-slating** membranes (still traditionally called "felt" although this material is no longer used) are breathable, allowing moisture vapour through but keeping water out. This helps avoid condensation in the loft.

At the apex, tiled roofs are normally capped with a purpose-made ridge tile, shaped to fit over the ridge board (Figure 4.7). A common pattern is **half round**, about 450mm long, bedded and jointed in a hard mortar. A similar capping can be used for the hip, but **third round** as a half round tile would sit too high on the shallower pitch involved (Figure 4.8). Hips to plain tiled roofs are sometimes formed with bonnets – purpose-made individual tiles (resembling a traditional bonnet and shown in Figure 4.1) fitted over the hip and interleaved with the general tiling. Valleys between roof slopes may be boarded with plywood and lined with lead or similar sheet material, with the tiling cut to the profile (slope) of the valley. However, in our example, interlocking purpose-made valley "trough" tiles are specified, which also interleave with the general tiling (Figures 4.1, 4.5). At the **eaves**, the tiles need to be started with a course of shorter tiles (165 × 165 × 15mm) to allow for the minimum of a double lap at the edge – this is traditionally called a "double course" of eaves tiles. Associated with the eaves are the **eaves fascia** (boarding facing up the ends of the rafters and providing a fixing for the gutters), a continuous plastic **eaves ventilator** (providing ventilation to the roof space) and **eaves soffit** (boarding providing an underside to the rafters). Finally, measured with the eaves are the rainwater goods – gutters, rainwater pipes and connection to the rainwater drains. Once the painting to eaves boarding and rainwater goods is included, there are four trades (or work sections) associated with pitched roof coverings, tiler, carpenter/joiner, plumber and painter. As with other sections, for details of the science and technology underlying the review here, specialist construction technology texts should be consulted.

Figure 4.7 Half round ridge tile

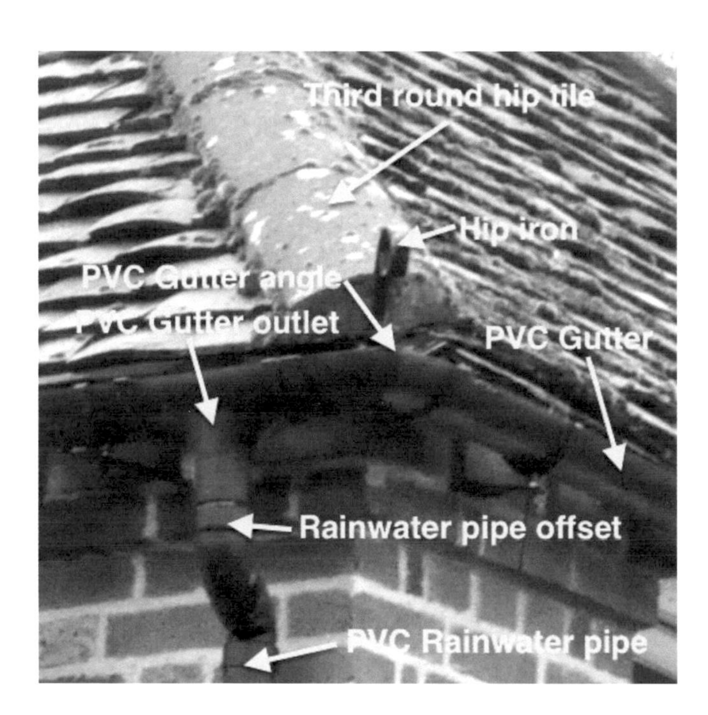

Figure 4.8 Third round hip tile

AA29 starts with an item for the tiling to the main roof. The dimensions have already been calculated – the length of the roof is 12400mm, and the slope length from end to end of the rafters is 4.50mm. Although the roof has hipped ends, this can be ignored for taking off purposes – geometrically the area is the same for a hipped as gabled roof because the pitch (angle) of the roof at the hips is the same as for the main roof. The area is timesed (multiplied) by 2, as there are two slopes to the roof. The area of each slope to the small roofs over bedroom 1 is added in, and, again, both the hipped ends and cut-ins to the main roof can be ignored – the hipped ends have the same area as if the roofs were gabled, and the cut-in sections have been measured with the main roof.

The description follows the specification for the roofing and the requirements of NRM2 18.1.1.1. The Work Section is *Tile and slate roof and wall coverings*, with *Plain tiling* as the main sub-heading and *Roof coverings* as the first subdivision. Thereafter, level 1 requires that the pitch (angle of the roof) be stated and level 2 that underlays and battens be included in the item (even though tiling and fixing battens are operationally two distinct activities). The general rules related to section 18 also require that the type, quality and size of materials be stated, along with method of fixing (of tiles and battens), minimum lap of the tiles and spacing (**gauge**) of battens.

The ridge tiles, hip tiles and double course of tiles at the eaves are all classed as *boundary work* at NRM2 18.3 and follow the categorisation under that sub-section. The notes to the sub-section give helpful guidance, including a definition of what boundary work means – "work associated with closing off or finishing off tile or slate roofing at the external perimeter, at the abutment with different materials or the perimeter of openings and voids". The notes also state that the work includes all rough and fair cutting (i.e., cutting the tiles to the angle of the boundary, which would involve a considerable cost for valleys and hips) and all other work in association with forming valleys. The latter could include plywood or timber linings and tilting battens, but not metal linings, which are expressly excluded. Note that when NRM2 says that certain work is *included*, it does *not* mean that an allowance in cost will not be made for it – it simply makes clear to the estimator that the taker off has not measured the work separately and the estimator must price for it in the item. Where work is expressly described as being *excluded*, it means that the taker off will measure a separate item for the work and the estimator does not price for it in the item. As with roof coverings, boundary work requires that the general rules to section 18 be followed, including specifying methods of fixing and the mix of mortar for bedding ridge and hip tiles.

The dimensions for most boundary work have been previously calculated in the roof construction, so figures just need to be brought forward to the relevant items. Ridge capping tiles are the same length as the ridge board, and NRM2 requirements are at 18.3.1.3.1. Although NRM2 requires "dimensioned descriptions" for boundary items, reference is usually made to manufacturer's literature or national or international standards containing standardised details and dimensions. NRM2 requires that the **plane** of the work be stated (horizontal, sloping, vertical, etc.), as this will affect the method of fixing and consequently the cost.

The hip capping tiles have been measured in the same way that the hip rafters were measured to the main roof (as though all four run from ridge to the corner of the building). This means that only the two *inner* hip coverings to the small roofs are measured. This is illustrated in Figure 4.9.

Immediately after the taking off for the hip tiles, on AA30 are measured six hip irons. These metal fittings (Figures 4.10 and 4.8) are attached by the tiler to the bottom of the hip rafter in order to stop the third-round hip tiles sliding off the roof over time. They are specified to be *screwed* to the hip rafter. In general, if a method of fixing is not specified for any item, it can be fixed by the most economical method (providing the fixing is fit for purpose). This will normally mean fixings to timber will be with steel nails.

For the hip irons, screwing is specifically required and must be identified in the item. If a particular type of screw other than steel is required (such as more expensive brass or zinc plated), then this further information must be given. The nature of the material to take the fixing is also cost significant, so NRM2 requires that the nature of the **base** (softwood timber) be stated in the description (NRM2 18.4.1.4.1).

The hip irons are painted with an appropriate primer and three coats of oil paint. This will normally consist of two undercoats and one finishing coat. The detailed specification for painting, including preparation and sanding between coats, is normally contained in a general specification or by reference to national/international standards. For small isolated items of less than 1m^2 each in area, such as the hip irons, NRM2 requires that the painting be enumerated (NRM2 29.3.3.2). One important variable associated with painting is whether it is *internal* or *external*, the latter

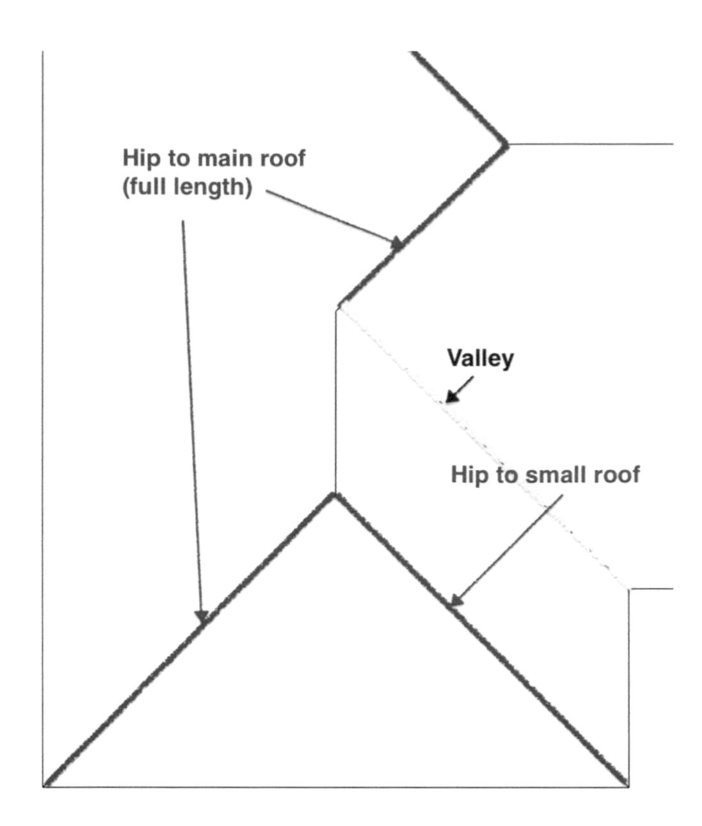

Figure 4.9 Hips and valley at small roof

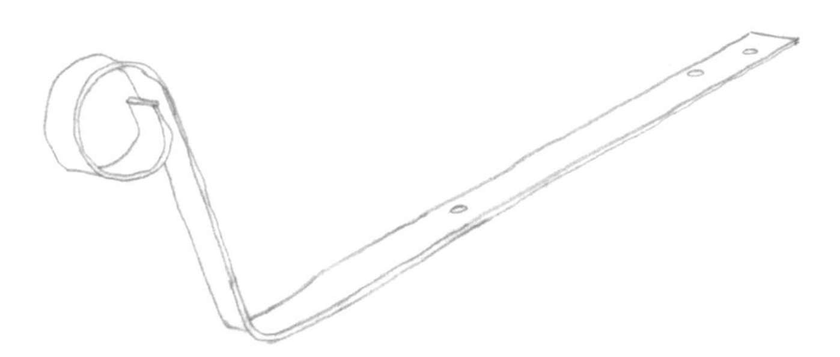

Figure 4.10 Hip iron screwed to softwood

being harder to access and subject to the weather. NRM2 requires, therefore, that internal and external work be given separately.

Work to the valleys is measured in the same way as hips, giving details of the construction and the length of the valley. In this example, the valleys are formed with purpose-made valley tiles interleaved with the general tiling (Figures 4.1, 4.5, 4.7). As the valley tiles are supported by the same battens as the general tiling, no further support is needed. To calculate the length for the "double" course of eaves tiles, the timber fascia and soffit, the taker off has girthed up from the external dimensions of the roof coverings 12400 × 8400mm and added 3000mm for the re-entrant area to give 44600mm. The double course is similar to and measured in the same way as the other roofing boundary items.

"Anded" on to the double course is the timber fascia measured in accordance with NRM2 16.4.1.1.1+3.1. NRM2 16.4 includes "fascias, bargeboards, soffits, etc.", and 16.4.1 requires that boards less than 600mm wide be given separately and the *finished* width and thickness stated. This means that the finished size of 150 × 25mm is as actually fixed and will be planed from a larger timber (allowing a 3-mm planing margin, at least 156 × 31mm and probably larger, as this does not match a standard sawn timber section). The level 2 descriptor requires that the plane of the item be stated (horizontal, vertical, etc.), and the level 3 descriptor requires that the finish be stated unless sawn – in this case, the timber is wrought (i.e., planed smooth). As the fascia is grooved to take the eaves soffit, the profile is also required at level 3 and is simply identified as being "once grooved". Cutting a groove or other indent in a piece of timber is traditionally called a **labour**, and there is a wide variety of labours worked on timber in forming such items as door and window frames, skirtings, architraves and window boards. For these items, each type of labour must be identified in accordance with section 22 of NRM2 (for more on labours to timber, see the Internal Finishings section). The final item in this column is for the continuous plastic eaves ventilator, measured under the same NRM2 rule as the double course of tiles.

AA31 continues with the eaves soffit (soffit is a traditional term for underside). Note that this is measured to its extreme girth and not on the centre line. As mentioned in regard to roof construction, forming angles (whether mitred or jointed) means that the timber must be cut to and is therefore measured to the outside edge. In NRM2 terms, the item is similar to the eaves fascia. A full explanation of the categorisation for this item, illustrating exactly how NRM2 works, is as show in Table 4.1.

The next item is painting to the eaves, and, as the fascia and soffit are a single item of decoration, the overall girth of the two components is calculated – 150mm for the fascia and 200mm for the soffit, making 350mm girth. The length of the painted area is taken as the same as the roof edge, involving a slight over-measurement (it should be measured on the centre line), but the girth (width wise) of the paint has been under-measured, as the thickness of the fascia has not been allowed for in the girthing. On balance, the slight over- and under-measurement should be about the same. The specification is for traditional oil painting involving five operations:

1 Knotting – treating any knots in the timber with a sealant that will not be dissolved by the sap in the timber.
2 Priming – applying a primer/sealer (often acrylic based) to the timber that will adhere well and form a base for the paint.

Table 4.1 NRM2 categorisation for timber eaves soffit

Section reference	
16	Work section 16, Carpentry
16.4	Fascias, bargeboards, soffits, etc.
16.4.1	Not exceeding 600mm wide, width and thickness 200 × 15mm
16.4.1.1	Plane of the work – horizontal
16.4.1.1.1	Finish as exterior-quality plywood
16.4.1.1.1.1	Location of the work, as eaves soffit
General rules	**"Information that should be provided"** • 5 "method of fixing where not at the discretion of the contractor", fixing by screws is specified • 8 nature of base – screwed to softwood **"Works and materials included"** • 4 "all work fixed by nails unless otherwise stated", so fixing the soffit with screws is stated

3 Stopping – filling all nail holes, joints and imperfections with wood filler.
4 Painting undercoats – specified by number of coats.
5 Painting topcoats – also specified by number of coats.

Included in general specification items will be requirements for treating the paint between coats and the detailed specification for the paint, often referring to a manufacturer's reference, national and international standards.

NRM2 deals with painting at Section 29, Decoration, and eaves are considered a General Surface at 29.1. The level 1 descriptor requires that work less than 300mm girth (width-wise, not lengthwise!) be measured in linear metres and, if over 300mm, girth in square metres. The level 2 descriptor distinguishes between internal and external work for reasons previously mentioned (internal work is less subject to the weather and easier to carry out).

The final set of items on AA31 and AA32 relate to the rainwater goods and provide an introduction to simple plumbing. For modern housing, rainwater goods are predominantly of self-finished PVC and consist of gutters, rainwater pipes, associated ancillaries (such as angles and outlets) and fittings (such as bends and junctions). So, all work is measured either in linear metres or as enumerated items. The side calculation at the bottom of the first column on AA31 is for the external girth of the gutter. As with timber, it is better to take the external girth rather than the centre line to allow for cutting at the corners. The length of guttering all-round the building is, therefore, calculated at 45400mm, and the item is covered in Work Section 33 of NRM2 – Drainage above ground. The general notes to the section divide the work into two main sub-sections of rainwater installations and foul drainage installations. Information to be provided includes the location of the installation and nature of the **background**. The location of gutters and rainwater pipes is fixed by implication (at the eaves as shown on drawings), but the nature of the background requires more explanation. For fixing items including rainwater goods, the material to which items are fixed has an important bearing on the cost of fixing. For example, it is much cheaper to screw to softwood timber than to hard masonry (which will need to be plugged). However, there can be a very wide range of different materials to fix to, and, for some items, NRM2 rationalises the range of materials into six "Backgrounds". These are listed on page 45 of NRM2 as:

- timber (including all types of hard and soft building boards)
- plastics
- masonry (includes brick, concrete, block, natural and reconstituted stone)
- metal, of any type
- metal-faced timber or plastics
- vulnerable materials (includes glass, marble, mosaic, ceramics, tiled finishes, material finishes, etc.)

Where NRM2 requires that the *background* be stated, then it can be slotted into one of these six categories. However, if the *base* is required to be stated, then the exact material providing the fixing should be specified. For some operations, such as nailing to softwood, omitting to state either the base or the background will not be a problem, as this is the most straightforward and cheapest method of fixing. For all other fixings, either by type of fixing (such as screws, bolts, clips, etc.) or nature of the base, the exact requirements must be included in the item.

General information to be provided in Work Section 33 also includes the method of spacing and fixing (gutters or pipes), description of supports and type of brackets or supports. These details are typically dealt with by referring to manufacturer's information or national or international standards and do not need to be included in the measured item. So, the gutter item on AA31 simply refers to specification item R10.10, which will include all these details. The framework of the item does, however, exactly follow the requirements of NRM2.

Following the item for the gutter are the ancillaries, consisting of pre-formed internal and external angles at the corners of the roof, the running outlets (to discharge water to the pipes) and balloon gratings at the top of the outlets. These are measured over the gutters, meaning that where an ancillary displaces a short length of gutter, no deduction is made to the guttering.

AA32 covers the taking off for rainwater pipes, with the side calculation working out the length of the pipe from gutter to drainage connection. The position of the rainwater pipes is not noted on the drawings,[3] so an assumption has been made that there are six, and this is stated on the taking off. Standard "house type" drawings produced by volume builders typically do not include rainwater pipe positions, and reference needs to be made to drainage layout drawings. In this instance, once layout drawings are available, the number of rainwater pipes will be altered if necessary. The NRM2 categorisation for rainwater pipes is exactly the same as the gutters, but the specification is different. Note that the background is classed as masonry and the pipe brackets are to be fixed with brass screws.

Following on from the rainwater pipes are the fittings. A plumbing fitting is a small item that adapts a pipe in some way – for example, a bend or elbow changing the direction of the pipe, a junction at which three or four pipes are joined or a reducer changing the diameter of the pipe. Pipes also need joining in their "running" length by straight connectors, but these are not normally measured, as they are needed at regular intervals and can be allowed for by the estimator. Fittings are different from ancillaries in that the latter are normally more complex items carrying out a distinct function separate from managing the normal flow of liquid. A tap connector, drainage valve, or gutter angle are examples of ancillaries. Measurement of fittings is, like ancillaries, "extra over" the pipes in which they occur – there is no deduction on the pipe for the length displaced by the fitting. As there is little difference in price between many fittings, NRM2 simply requires that they be differentiated by size (NRM2 33.3.1 – less than or greater than 65mm) and described by the number of ends the fitting has (thus a bend has two ends and a T junction has three ends). In our case the taking off is for two offsets (shallow angle bends as seen in Figure 4.8) with two ends (NRM2 33.3.1.2). Extra over items are linked to the main measured item (the rainwater pipe), and reference is made to a separate specification, so the exact nature and size of the fitting will be apparent. The purpose of the two bends is to bring the rainwater pipe from the edge of the roof below the gutter in to the face of the external wall.

The final measured item is for the connector from PVC rainwater pipe to the vitrified clay drain. This is a pipework ancillary measured in accordance with NRM2 33.2.1. Again, reference is made to a specification where full details of the nature of this item will be found.

The last two items in this section are for marking out and testing the rainwater installation. The first is related to attendance on the specialists carrying out the installation – in this case, plumbers. The plumbers will **mark out** the routes of pipes and position of associated ancillaries but will not carry out the work in cutting into and through walls, floors and other elements. This work, commonly known as "builder's work in connection" with the installation, is executed by other specialists or trades. The estimator is given the opportunity to price for the cost of the marking out. The fact that, especially for small works, plumbers often do their own associated builder's work means that this item is frequently not priced. The second item relates to the need to test service installations before completion. For some services, this is an important provision, as timely testing and commissioning before pipes, conduits and cables are covered up saves much remedial work later. Accordingly, the estimator is given an express opportunity to price for testing the installation.

Key points covered in this chapter:

- Outline of the history and technology of traditional roof coverings in the United Kingdom – plain tiles, interlocking tiles, slates and stones
- Names and functions of features to plain tiled roofs, including ridge capping, bonnet and third-round hip tiles, hip irons, lined and interlocking tile valleys and eaves tiles
- Timber eaves details and PVC rainwater goods
- Identifying items for taking off roof coverings
- Constructing item descriptions for roof coverings using NRM2
- Applying explanations to the example

Notes

1. Statute Edward IV, 17 C4 1477.
2. Cotswold stone roofs are traditionally laid to diminishing courses, with larger stones set on the lower courses and the size of stone and **gauge** (spacing of supporting battens) reducing as the top is approached.
3. The latest drawing revision in Appendix 3 has the rainwater pipe positions marked on plan and elevations.

5 First-floor construction

AA33 and AA34 show the taking off for the first-floor construction. Unlike the previous sections, drawn constructional details are limited, and what details there are must be supplemented by reading the specification. Some information can be obtained from the sections – for example, the direction of span of joists and the position of the stairwell – but the details shown in Figures 5.4 and 5.5 are based on the taker off's knowledge of technology as applied to the specification. As before, it is important to understand the technology before tackling the taking off.

Suspended upper floors for domestic construction in the United Kingdom are predominantly of timber joists and boards. This is generally only the case for single-unit housing and not for apartments. For the latter, where fire protection between dwellings is stringent, concrete floor construction is the norm. Timber construction consists primarily of two elements – a floor covering (Figure 5.1) and the supporting joists. Floor coverings were until recently exclusively timber boards, initially plain edged rectangular boards of about 150 × 25mm finished section. Plain edged boarding is versatile, as it is easy to raise single boards to give access to services. However, as each board is acting separately in resisting point loading, it is less efficient in distributing loads than more modern tongued and grooved boards. Using tongued and grooved boards (Figure 5.2) permits a thinner and narrower section (often no more than 119 × 19mm finished section) at lower cost. The downside of this type of boarding is that, in order to remove single boards, the tongue needs to be cut through. Once a board has been weakened in this way, it will often deflect more and give a characteristic and irritating squeak. Tongued and grooved boarding has now largely been supplanted by high-density particle boarding. This boarding is structurally efficient, more stable and cheaper than timber boarding but is even harder to adapt if access is needed to the floor void. Particle boarding may disintegrate if allowed to get wet, so a moisture-resistant grade is needed in vulnerable areas (e.g., kitchens and bathrooms). Nail fixings are also prone to loosen over time, causing the flooring to squeak when walked over, so it is recommended that the flooring be screwed down.

Supporting joists are traditionally softwood timber sections fixed at 400-mm centres, but increasingly composite lattice or web beams are being used for volume housing (Figure 5.3). These are factory made and delivered as units to fit the house type involved in much the same way as roof trusses. For measurement, as composite units, they are enumerated in accordance with NRM2 16.2.1.5.1, giving a manufacturer's reference. Traditional joists need strengthening if they are to support heavy partitions (even if not loadbearing) and also for **trimming** around openings, for example, for a staircase (see Figure 5.4). To support partitions, a common method of strengthening is to "double up" the joists – that is, nail two standard joists together. This is preferable to increasing the section of the timber used (e.g., from 50 × 100mm to 75 × 100mm), as it means all joists of the same size can be ordered. For trimming an opening, it is now also usual to double up the standard joist for the trimmer and trimming joists.[1] All junctions between **trimmed** and **trimmer**, trimmer and **trimming** joists are now made using steel "joist hangars", although in the past intricate "tusk tenon" joints may have been used.

DOI: 12.01/9781003253129-6

Figure 5.1 Timber floor boarding

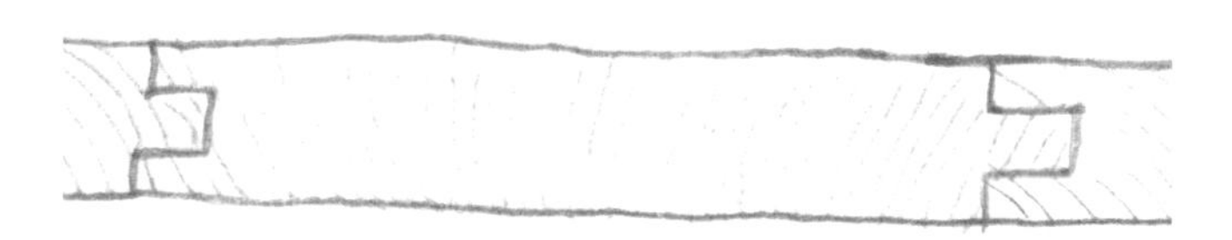

Figure 5.2 Tongued and grooved boarding

Figure 5.3 Composite floor joists

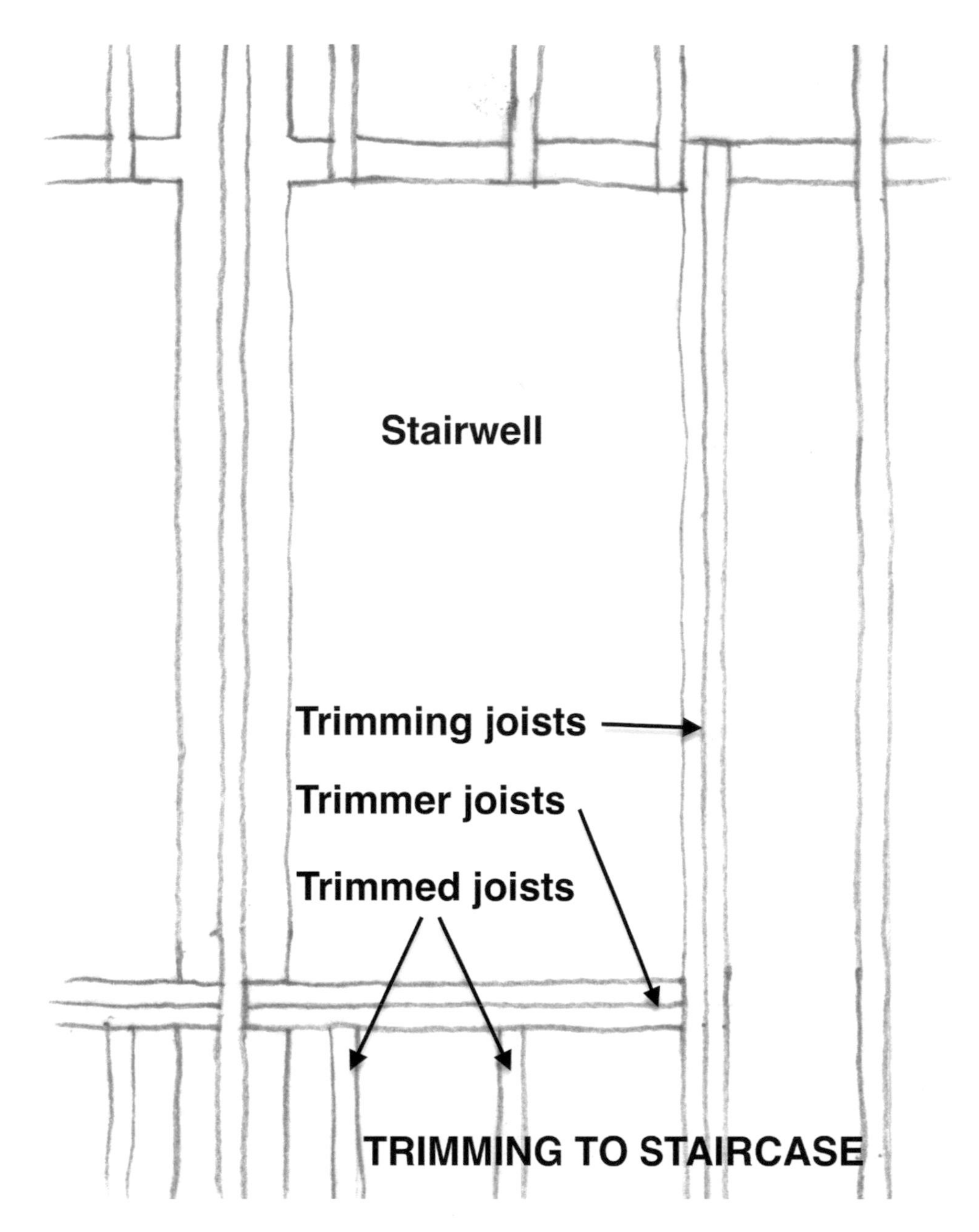

Figure 5.4 Trimming to staircase opening

Timber upper floor joists are of a "deep" section – that is, the ratio between depth and width of the joist is high – and this means they may deflect sideways when loaded. To limit this, it is usual to provide resistance to sideways deflection in the form of "strutting". This is occasionally "solid", consisting of full-size timber joist sections wedged in between the joist in rows about 2m apart. However, solid strutting uses a lot of material and blocks access for services, and it is difficult to tighten the struts to provide rigid lateral resistance. Short galvanised mild steel bars, or small section timber struts (about 38 × 38mm or 50 × 50mm) fixed in a "herringbone" pattern are more economical of materials and easier to wedge in between the floor joists (Figure 5.5). If deflection is expected in one direction, then the strutting will be in a "single" pattern (all the struts going in the same direction). For domestic floors, "double" **herringbone strutting** is more usual, designed to provide resistance in the two directions perpendicular to the span of joists. In our example, a row of double herringbone timber strutting is specified for each span of floor joists – that is, two rows per dwelling. As with other sections of measurement, for more detail on the science and technology of suspended domestic floors, specialist texts should be consulted.

AA33 starts by listing the drawings and providing a brief taking-off list. The first item is for the floor boarding, measured in m^2. The taker off has measured over the whole internal area of the two dwellings, deducting for the re-entrant area, stair opening, party wall and loadbearing partitions and taking the dimensions from the substructure oversite measurement at AA11. The description is based on NRM2 16.4.2.1.1.1.1 plus the brief specification. Where the staircase opening is formed, there is an exposed edge of flooring, and this is finished with a "nosing" (Figure 5.6) – a small rounded timber trim of the same depth as and glued to the flooring. This is measured in accordance with NRM2 22.3.1.2, General joinery, Cover fillets, stops, trims, beads, nosings, etc., giving the overall cross-sectional area and the number/type of labours. In this instance, the labours consist of rounding the front edge and **tonguing** the back edge to a groove in the adjacent floor boarding.

In the following waste calculations on AA33, the number of floor joists are first calculated (note that the direction of span of the floor joists is not obvious from the plans and must be deduced as being from N to S from the sections). The overall internal width of a single dwelling is taken and divided by the centres (spacing) of the floor joists (400mm). The answer gives a remainder, so an additional joist is added and a further joist to allow for one at the end. Finally, a further joist is added to allow for "doubling up" the joists for the non-loadbearing partition between bedroom 1/2 and bathroom/bedroom 3. The length of the joists is calculated by taking the internal depth of the dwelling of 7350mm, adding for two end bearings (the joists are built into the external walls) and for a lap over the loadbearing partitions. Note that, although the overall length of joist taken is 7650mm, this is not in one length, as they are jointed over the partition, with a simple lap.

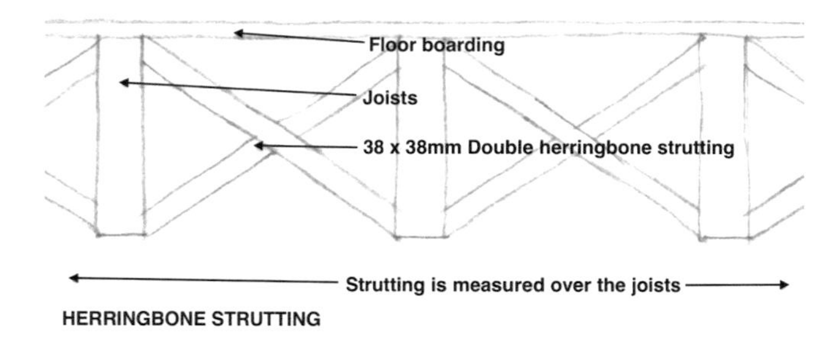

Figure 5.5 Double herringbone strutting

NOSING TO BOARDING OR STAIR TREAD

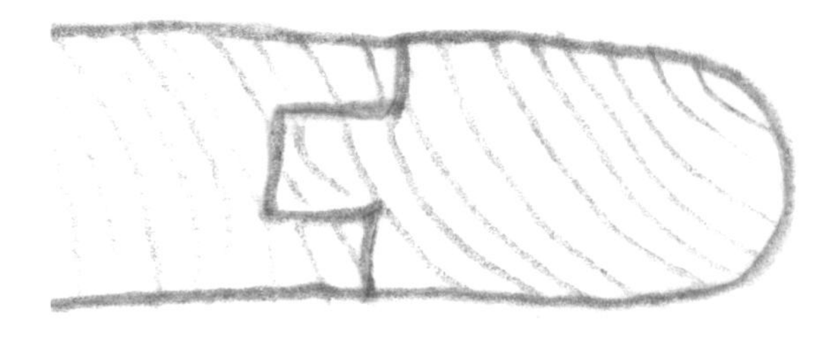

Figure 5.6 Nosing to boarding

The joists are shorter where the front wall comes in at the hallway/entrance area (the re-entrant). This re-entrant is 2100mm wide, and there are 2100/400 = 5 shorter joists here. The joists are also shorter where "trimmed" for the staircase opening. The opening is 900mm wide, giving 900/400 = 2 shorter joists here. These lengths of joist are deducted from the total length measured. The final part of the waste calculation on AA33 and the top of AA34 is for the additional trimming and trimmer joists to the side of the staircase. The trimming joist spans across bedroom 3, the partition and staircase opening, and the trimmer joist spans the width of the stair opening plus a bearing into the party wall.

Having calculated the lengths of all the floor joists, trimming joists, trimmer joists and extra joists to support the partitions, including allowing deductions for shorter joists at the re-entrant and over the stairwell, it remains to write the item on AA34. Note the method of working – as with other sections, the taker off has grouped all the waste calculations together at the start and rationalised the dimensions into a single item with all the dimensions booked in one column. This approach is more efficient than measuring items piecemeal – by, for example, measuring and booking the gross lengths, then adding a further item for trimmers and trimming joists and then writing yet further items for the partition support joists and the deductions.

The floor joist item is the same as for the ceiling joists measured with the roof construction, except that the section size is different and the joists are described as being built in. The sign-posts to the right of the item make clear what has been measured where. For example, the main joists amount to a length of 7.65m multiplied by 16 for each dwelling and multiplied by 2 for the two dwellings in the block. The trimming joists are 3.60m, long and they are multiplied by 2, as there are two nailed together. They are multiplied by 2 again for the two dwellings in the block.

The herringbone strutting is measured as a simple length over the joists (i.e., the actual lengths of the small section timbers forming the struts are not measured). The E–W width of one dwelling was calculated as 5550mm on AA33 in working out the number of floor joists. As there are two rows of strutting, this is multiplied by 2 in the dimensions, then by 2 again for the two dwellings in the block. There is no strutting where the stairwell is formed, so 0.90m is deducted for this to both dwellings. The star (*) and note at the bottom of the column make clear that the taker off has based the measurement on two spans per dwelling.

The final two items are for the carpenter's metalwork for the connections between trimmed and trimmer joist and trimmer and trimming joist. For the former, a 225 × 50–mm **joist hanger** is used to match the size of the joist, and for the latter, a 225 × 100–mm hanger is used to match the size of the two joists nailed together in forming the trimmer. The hangers are designed to form the connection without projecting above or below the surface of the joists (so that flooring or ceiling finish is not affected) and are generally fixed using nails. NRM2 groups metal fixings, fastenings and fittings at 16.6 with a requirement for a dimensioned description 16.6.1 (usually given by a manufacturer's reference – e.g., Simpson Strong-Tie UK, SAE500/50/2), the type of fitting (in this case joist hanger – 16.6.1.7) and the method of fixing (16.6.1.7.1). Note that the item does not give the method of fixing, but general rules at the start of NRM2 section 16 under "works and materials included" deem that "all work is to be fixed by nails unless otherwise stated". The "nature of base" is also required in the general rules for carpentry items, but, as fixing to softwood is the least expensive option, omitting to state the base will not be a problem. A general heading in the Carpentry section of the Bills of Quantities stating that all fixings are to softwood timber unless otherwise stated would also deal with the omission. The nature of fixing may also be clear from the manufacturer's specification.

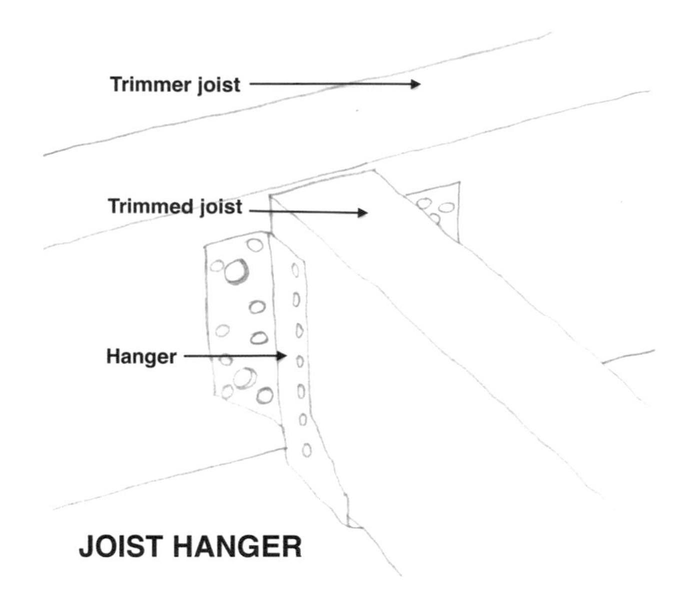

Figure 5.7 Steel joist hanger

Key points covered in this chapter:

- Outline of the technology of suspended timber floors in the United Kingdom, including traditional and modern types of boarding, joists/beams and strutting
- Supporting partitions and trimming openings, including defining terms. The purpose and use of joist hangers
- Identifying items for taking off suspended timber floors
- Constructing item descriptions for suspended timber floors using NRM2
- Applying explanations to the example

Note

1 Trimmer joists support trimmed joists – those joists that have been cut short to form an opening. Trimming joists support the point load(s) of trimmer joist(s) as well as performing the role of a standard joist. See Figure 5.4.

6 Internal finishings – ceilings, floors and walls

Internal finishings were touched on in the introduction, where the main work to the walls internally was used as an example for the form of taking off. This is now expanded to include further details for walls and work to floors and ceilings. Finishings are generally taken to include internal non-structural work to these surfaces but exclude work that is closely associated with other elements. So, skirtings are measured with the wall finishings, but window boards are measured with windows and doors; ground floor boarding is measured as a floor finishing, but upper floor boarding is measured as part of the floor structure. In the traditional construction sequence, internal finishing work is a key activity separating joinery, electrical and plumbing **first fixings** from their corresponding **second fixings**. Once finishings are complete, the building, overall, is approaching completion. Taking off internal finishings and associated boundary items gives a good overview of the building and generates a lot of dimensions. It is, therefore, a good section for inexperienced practitioners to gain proficiency.

As with other sections in this book, it is useful to consider the technology of finishings before tackling measurement. Taking ceilings first, until the middle of the 20th century, site-applied "lath and plaster" was the predominant finish, and this technology is still frequently encountered in repair and restoration work. **Laths** are thin strips of timber nailed closely together to the joists to form a **key** for the plaster. When plaster is pressed into the narrow slots between laths, it spreads outwards and sets, fixing the plaster in place (Figure 6.1). The plaster is lime based (calcium hydroxide) mixed with fine sand and animal hair and applied in two or three coats. Lime plaster takes a long time to dry out, and surfaces would, therefore, be left unpainted for a correspondingly

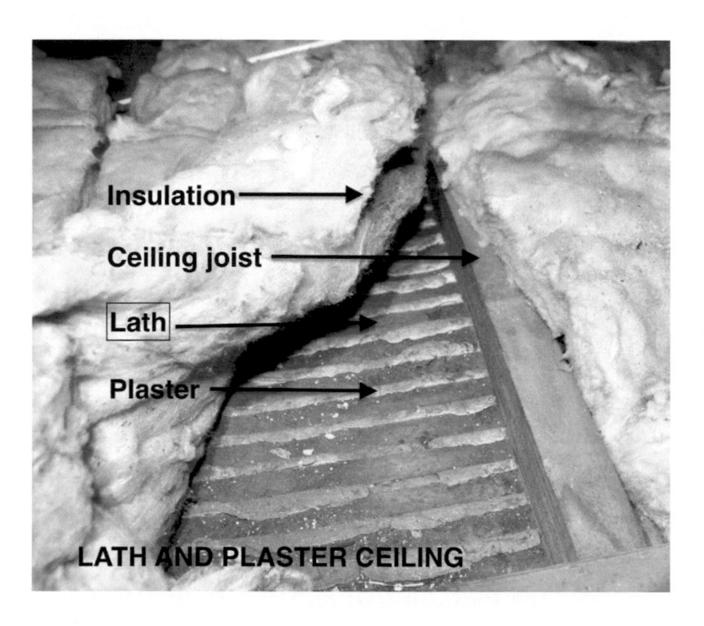

Figure 6.1 Lath and plaster ceiling

DOI: 12.01/9781003253129-7

long time. Using lime is also hazardous, and, as a desiccant, can leave tradespeople rather thirsty! The early part of the 20th century involved experimentation with several board materials, including vegetable fibre and asbestos cement. However, the invention of plasterboard (precast plaster sandwiched between two sheets of paper), which is better and safer than these, quickly supplanted their use. Now, nearly all domestic ceilings are plasterboard, either designed for direct painting or to take a thin "skim" coat of site-applied plaster. The plaster in plasterboard sheets is gypsum based, which sets hard and is suitable for pre-casting. Plasterboard is commonly 9 or 12.5mm thick and needs supporting on all edges – on the long edge by floor or ceiling joists and across the short edge by inserting **noggings** (small section timber battens). Boards are nailed to joists and noggings with broad-headed galvanised nails, and the work is usually carried out by specialist "tackers". The skim coat plaster (where used) is also gypsum based, and about 3–5mm is applied by skilled plasterers to even out imperfections and mask joints in the plasterboard.

Domestic floors were traditionally timber board and joist on both ground and upper floors. However, for ground floors, suspended timber is now unusual, having been supplanted by suspended concrete or ground-bearing concrete beds as in our example. In modern work, the floor will be insulated either under or over the concrete. In our example, the insulation is placed over the concrete and is, therefore, measured with the internal finishings and not with the structure. Solid concrete ground floors are often finished with a cement and sand **screed** bed (varying from about 40–75mm thick), with a top surface treated to suit the final covering – for example, for smooth vinyl, by **trowelling**. Screed beds are prone to cracking if laid on a soft insulation, so they are not suitable in our example, where the insulation is laid over the concrete and below the flooring. An alternative, commonly used by volume house builders and employed in our example, is to lay flooring-grade particle board directly over the insulation as a **floating floor**, without fixing it to the concrete. The flooring is held in place by the skirtings and its own self weight.

In our example, the **skirtings** have been measured with the floor finishes. Skirtings are a joinery second fixing. Unlike floor boarding and internal **door linings**, they will be fixed *after* the walls have been plastered. Skirtings and similar items of joinery (window boards, architraves, door linings, etc.) are usually specified to a standard section shape, sometimes based on architectural profiles taken from classical masonry – **ovolo** (Figure 6.2), **taurus** and **ogee**, for example. In specifying using these terms, the estimator will know the work involved in producing the profile. For non-standard sections, the architect will need to draw the required profile (or select from a catalogue) for both pricing and construction purposes. In making both standard and non-standard items, the timber piece being worked will be run through a rebating machine capable of producing a large number of profiles. A complex machine may be able to handle a large number of different **labours** on the timber simultaneously, or it may be necessary to pass the timber through the machine several times using varying settings to get the required profiles. Figures 6.3 and 6.4 show an example of labours worked on timber to form a traditional window sill.

The finished section shown in Figure 6.3 is obtained from the original sawn section by first planing it to a wrought surface. Figure 6.4 shows eight labours worked on the timber to give the final profile. Each labour has a traditional name (**rebate**, **sunk weathering**, **groove**, etc.), which can often be linked back to their purpose – for example, a **drip** is intended to direct water away from a wall and to "drip" off the sill, and a **weathering** is a sloping rebate allowing water to run off the timber. In modern work there is no real difference in the type and price of each labour, as each is run through a machine. For pricing purposes (not manufacturing), the number of labours can simply be counted – eight in Figure 6.4. However, if a labour is to be **stopped**, as a sunk weathering may be before the ends of a sill, this is more expensive (the piece has to be taken from the machine without running right through) and should be described separately.

In our example, there are two specifications of skirting – clear finished hardwood and painted softwood. For the former, the timber must be kept clean for varnishing, and the skirting is to be

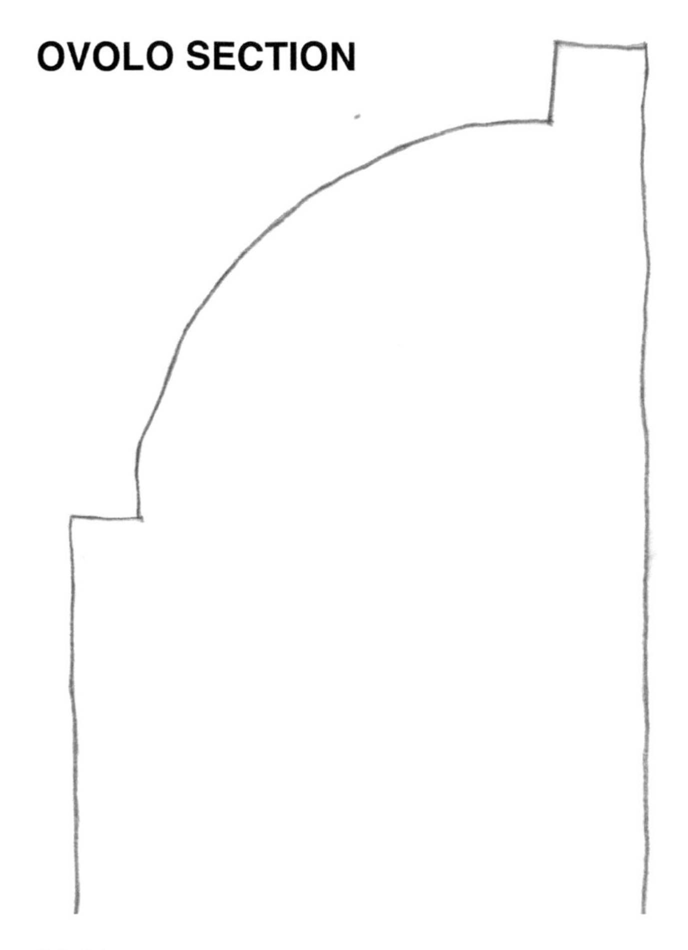

Figure 6.2 Ovolo section skirting

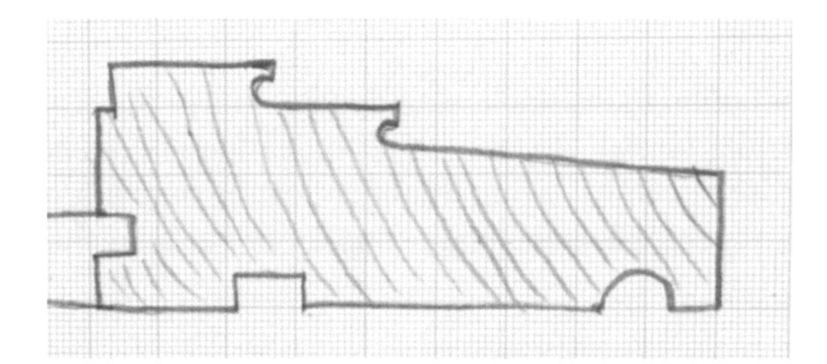

Figure 6.3 Section through timber sill

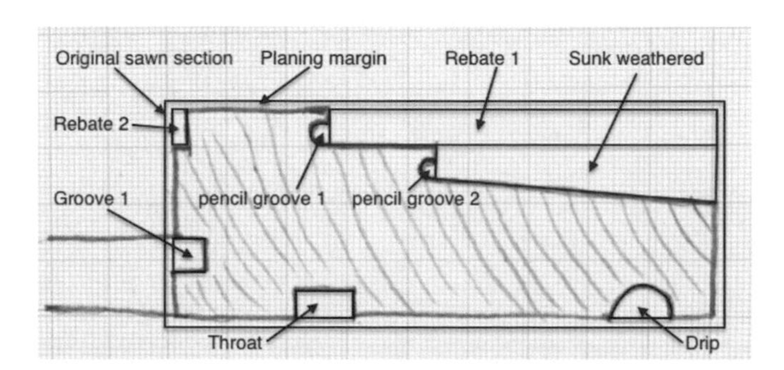

Figure 6.4 Labours on timber sill

fixed using screws and **pellets** – small cylindrical plugs of the same wood as the skirting let into the timber to cover the screw heads. For the latter, the skirting is to be painted, so less care need be taken in preparing the timber, and the screw holes can simply be filled with **stopping**.

The traditional wet plaster wall finish was described in the introduction to this book and is specified in this example. Where wet plaster is specified, it is almost invariably two coat render and set using gypsum-based plasters. For work in areas subject to dampness, cement and sand mortar sometimes supplants gypsum for the render coat, as it is more damp resistant, and for historic, work lime-based plaster is often more appropriate. Although wet plaster is still in common use, for volume house-building, developers prefer **dry lining**. Dry lining involves lining the external walls and loadbearing masonry partitions with plasterboard set on gypsum plaster **dabs**. The plasterboard is then **levelled** by pressing the sheet onto the wall face. The surface of the board is designed for direct painting, without the need for a skim coat of plaster, and ceiling plasterboard will also be set up in the same way. Non-loadbearing partitions are usually of composite plasterboard construction with surfaces also prepared for direct decoration. With the exception of the plaster dabs and a small amount of skimming plaster to cover board joints, dry lining completely eliminates wet plaster.

Most wall plaster is applied in broad areas, but working around windows and doors requires more attention. It is more difficult to apply plaster to narrow window reveals (the internal side of the window opening), and the plasterer has to run an **arris** (external corner) at the junction of reveal and main wall plaster. Modern arrises are not run freehand but are formed using metal angle beads (Figure 6.5) set onto the masonry wall with plaster dabs. The plaster is worked up to the bead, and the projecting rounded edge of the bead forms the angle. Plastering the top soffit reveal of an opening also presents the problem of encouraging the plaster to stick to the underside of the lintel. Steel lintels, as specified in our example, are supplied with slotted holes to the underside to provide a key for the plaster, but even with these, the plaster may be too heavy to stick well, and a lightweight plaster may be specified. Plaster will not stick to the edge of the timber plates to the tops of the internal skin of blockwork and loadbearing partitions. To provide a keyed surface here, an effective solution is to nail a strip of expanded metal lathing to the side of the plate (Figure 6.6).

As outlined in the introduction to the book, wall decorations in our example are predominantly emulsion paint applied to the plaster. A standard specification is for one thinned coat and three full coats. The exact requirements, including treatment of the bare plaster surfaces and preparation between coats, will be specified by the paint supplier or national and international standards.

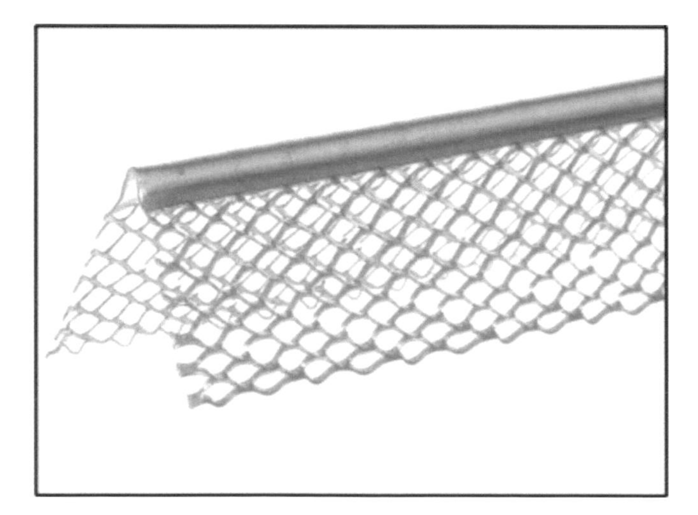

Figure 6.5 Metal angle bead

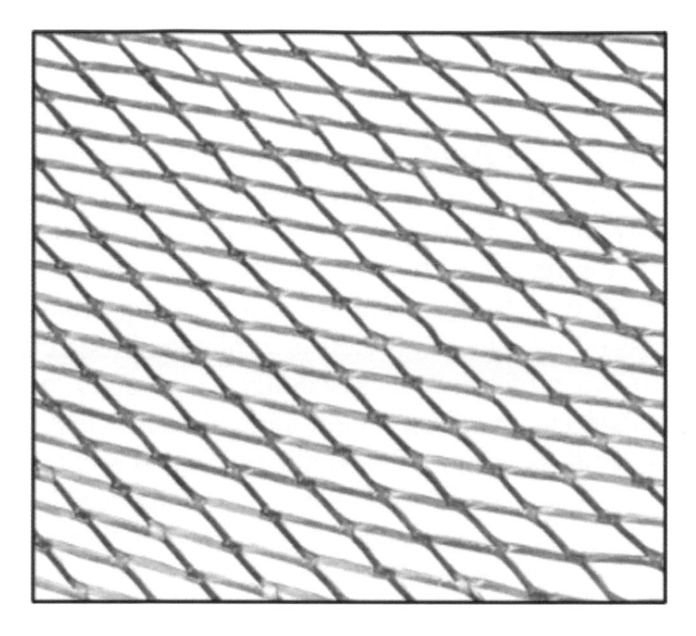

Figure 6.6 Expanded metal lath

For the lounge, dining room and hall, a wallpaper in place of emulsion is specified. As wallpaper fashions change rapidly, it is usual for an allowance to be made for supplying the paper as a **Prime Cost Sum** (for example, £60.00 per standard piece of paper – 10.00m × 600mm), with the estimator pricing this figure into the estimate. NRM2 page 10 gives a formal definition of Prime Cost Sum:

> A sum of money included in a unit rate to be expended on materials or goods from suppliers (e.g., supply-only ceramic wall tiles at £50/m2, supply-only door furniture at £120/door or supply-only facing bricks at £480/1,000). It is a supply-only rate for materials or goods where the precise quality of those materials and goods is unknown.

When the paper is purchased, the actual cost per piece will be substituted and the cost adjusted. Papering involves priming bare walls with size (glue), pattern matching and cutting to edges.

The bathrooms are fully tiled, and, as with papered rooms, a Prime Cost sum is specified for supply of the tiles. The estimator will price the tiling at a rate including this figure, and the sum will be adjusted when the actual tiles are purchased. In the kitchen, a small area of walling above the sink and adjacent units (called a **splashback**) is also to be tiled, and the supply of tiling is treated in a similar manner. Modern wall and floor ceramic tiles are glued directly to the base – plaster for walls and a cement and sand screed for floors – so no adjustment in taking off is necessary for the plaster or screed. However, for historic work, bedding tiles "wet" to an unset base was common. So, wall tiling would be fixed on an unset cement and sand render coat, and floor tiling would be fixed on unset bedding mortar to a rough screed. As with plaster, tiling fixed in narrow areas, such as window reveals, is more expensive. At changes in direction of tiling (for example, to an arris), a rounded-edge tile may be specified, and to any exposed corners, a corner tile rounded on two adjacent edges may be specified (an REX tile). The exposure of tiling to water and condensation is an important specification determinant, as tiles will **lift** off the base if not properly fixed. For areas of high exposure, such as showers, the tiles will be **bedded solid** – that is, without any air spaces in the adhesive. For areas of lower exposure, such as general bathroom and kitchen areas, **spot** or **comb bedding** at lower cost may be acceptable. Similarly, joints to tiling in wet areas will need to be filled solid with a fully waterproof grout, whereas this is not essential for general areas.

Painting to timber and metal surfaces was introduced previously in roof coverings, but the hardwood skirtings in living and dining rooms are treated with a synthetic varnish. This will involve ensuring the skirting is kept clean and dry, as water will indelibly stain timber. Then a preliminary staining coat and two coats of oil-based finish is a common specification. Stopping will be matched in colour to the timber, and pellets will be matched (by the joiner) to the grain of the timber or **grained** by the painter to give a matching grain effect. Increasing use is being made of water-based **acrylic paints** for all work, as it is more environmentally friendly, but in this example all paints are **oil based**. The treatment of timber for painted skirtings (knot, prime, stop and three coats of gloss oil paint) was previously explained in the chapter on roof coverings.

AA35–AA42 covers the taking off for finishings, starting, as with other sections, by listing the drawings used and producing a short taking-off list. Ceilings are measured first, and the waste calculation for these goes back to the overall footprint of the building of 12000 × 8000mm, reducing this by 2 × 325mm for the thickness of the external walls. This area is further reduced by the re-entrant at the front of the building, 4650 × 1500mm, and the party wall, 5850 × 250mm. As these figures have previously been calculated for the concrete and hardcore oversite, it is not really necessary to re-calculate them in this section. Ceilings will be cut between loadbearing partitions, which are erected before the upper floor and ceiling joists are inserted, but will be in place prior to erecting the non-loadbearing partitions. So a deduction is made for loadbearing partitions but not for non-loadbearing partitions (Figure 6.7). A note in the right-hand column of AA35 reminds the reader of this. The preceding waste calculation is for the lengths of the loadbearing partitions to be deducted from the area. Again, this length has been calculated in both the substructure and for the walling and partitions, and it is not strictly necessary to re-calculate it here. Note that no deduction has been made in the finishings section for the staircase opening – although to do this here is quite acceptable, it is dealt with later with the staircase in Chapter 9.

NRM2 Section 28, Floor, wall, ceiling and roof finishings, deals with ceilings at 28.9, work over 600mm wide at 28.9.2 and a requirement to state the base in the general rules to the section. The remainder of the item contains the specification – 12.5-mm plasterboard with a 3-mm skim coat of finish plaster all as the specification. General specification clauses and reference to manufacturer's information or national and international standards will amplify these brief requirements.

NRM2 Section 29, Decoration, deals with the emulsion paint; ceilings, as a general surface at 29.1; areas over 300mm wide at 29.1.2; internal work at 29.1.2.1; and a requirement to state the base to be painted in the general rules to the section (plastered ceiling). As with the ceiling finish, the painting item refers to a specification, which in turn will refer to detailed standards. Signposts against each dimension amplify what exactly has been measured where – first the main area for the building multiplied by

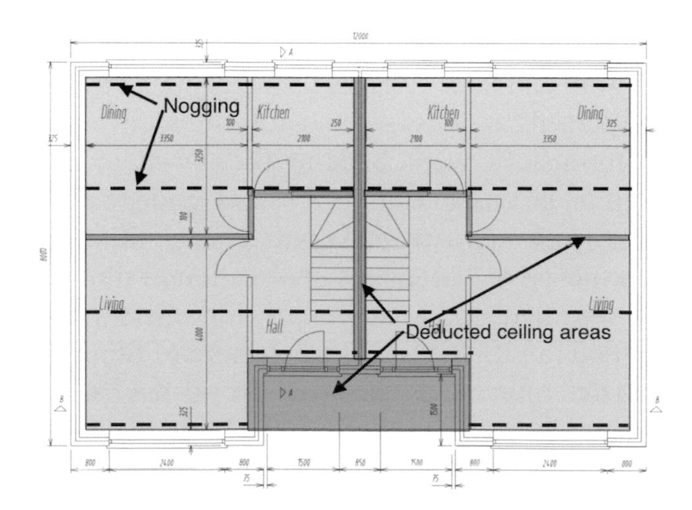

Figure 6.7 Nogging to plasterboard ceiling

2 for the two floors, followed by deductions clearly labelled for the re-entrant area, party wall and loadbearing partitions. The order of taking off follows the conservative convention to measure gross (i.e., overall) first and deduct afterwards. Note that, although decorations to ceilings are carried out after all the partitions are fixed, no deduction has been made for work above non-loadbearing partitions – the area painted is taken as the same area of plasterboard, giving rise to a slight over-measurement. In the context of the relatively low value of emulsion painting, this is insignificant.

For convenience in taking off, the insulation above first-floor ceilings is measured next, although this item is related to roof construction. Note that in the roof construction taking off, there is a To Take note for the item. NRM2 section 31 deals with insulation quilt at 31.3, requiring the thickness to be stated (250mm) at 31.3.1, laid across joists at 31.3.1.3 and horizontal at 31.3.1.3.1. The area taken for the quilt is to the outside of the external walls to allow it to lap over the walling and connect to the wall insulation. The areas of the re-entrant and party wall are deducted from this area.

AA36 starts by calculating the length of nogging fixed to the end of plasterboard sheets. Nogging is an historic term for filling between timber beams, but in the present context means short lengths of small section timber fixed between joists. A plasterboard sheet is typically 2400 × 1200mm and will be fixed at the long edges to the floor/ceiling joists (NB: as joists are usually at 400mm or 600mm centres, they conveniently suit the width of a sheet). The sheets in our example require support on all edges, and the nogging provides this on the short 1200-mm edges. The length of nogging will be the overall internal width of the building of 11350mm minus the thickness of the party wall at 250mm = 11100mm. The number of rows of nogging will be the internal depth of the building of 7350mm divided by the length of a plasterboard sheet, giving 3 plus one extra starting row. Note that no adjustment to the length of nogging has been made for the re-entrant area – so the length has been slightly over-measured. Nogging is not normally needed against partitions, as the wall plaster will sufficiently hold the plasterboard in place. Also, the plasterboard tackers will usually stagger the joints of the plasterboard, but this will not affect the overall quantity of nogging. NRM2 covers nogging at 16.3 as "backing and other first fix timbers"; 16.3.1 requires a dimensioned description (given as 50mm × 50mm nogging), 16.3.1.2 classifies them as "battens" (a general term for a backing timber), and the general rules to the section require that the nature of the base be given (softwood).

The next waste calculation is for the **coving** fixed between wall and ceiling finishes in the lounge, dining room, hall and kitchen (Figure 6.8). This is simply described as Gyproc Standard

Figure 6.8 Ceiling coving

coving in the specification, but, as with other items, the description is backed up by further details and a reference to manufacturer's information and national and international standards. To work out the lengths of coving, each room has been girthed up individually and entered in the dimensions. Both the waste calculation and the signposts to the item provide location information. NRM2 deals with coves at 28.17, with a requirement to state the girth and shape at 28.17.1, the plane of the work as horizontal at 28.17.1.2 and the method of fixing and nature of background at 28.17.1.2.1+2. A note is added that no deduction has been made for paint behind the coving, but, as the coving itself would be painted, this is a minimal over-measurement.

Following the coving is a note; "TT – Loft hatch with internal doors". This is a **To Take** note, reminding the taker off that the loft hatch has not been measured, nor any deduction made for the finishes.[1] As a general procedure, for items on the boundary of one or another taking-off section, or for items that might be forgotten if not measured in the expected sequence, it is a good idea to make a brief **To Take** (TT) or **Item to Measure** (ITM) note. The taker off is then reminded to measure the item before finishing all the taking off. This procedure is also commonly used when information for an item is missing or yet to be supplied, as it allows measurement to be deferred until it is available.

The next item is the ground floor finishes, consisting of a particle board floating decking on insulation. Unlike the ceilings, the floors have been measured room by room between both loadbearing and non-loadbearing partitions. This is because all partitions are in blockwork and cannot be built off a floating floor. The areas for each room have previously been calculated or can be directly taken off the drawings. Boarding is a carpentry item, measured in accordance with NRM2 16.4, boarding, etc.; over 600mm wide; (16.4.2), horizontal; (16.4.2.1), location of the work as floors; (16.4.2.1.*.1) and nature of base as the general rules to the section. Reference is made to the specification, which will again be backed up by general specifications, manufacturer's details and national and international standards. Thirty mm of sheet insulation is laid below the boarding, measured in accordance with NRM2 31, Insulation, fire stopping and fire protection; boards; (31.1), plain areas; (31.1.1), horizontal; (31.1.1.1). Two notes after the insulation item make clear the assumption that flooring is fixed between partitions and that the upper floor finish has not been measured, as it is with the floor structure measured previously.

Next on AA37 are two items for skirtings. The living and dining rooms have hardwood skirtings, plugged, screwed and pelleted to blockwork. All other skirtings are of painted softwood. NRM2 deals with skirtings at 22, general joinery, thence 22.1, skirting and picture rails. 22.1.1 requires that the dimensioned overall cross-section be stated – described in the item as 150 × 25mm (nominal) ovolo pattern. As mentioned, ovolo is a standard profile based on classical architectural sections, so the labours required at 22.1.2 are clear from the pattern – that is, the machining needed to finish the skirting to the size and shape required. NRM2 section 29 deals with the decoration to skirtings. Skirtings are classed as a general surface at 29.1; less than or equal to 300mm girth; (29.1.1), internally; (29.1.1.1) and the nature of the base is given as required by the general rules to the section. Thereafter, treatment to hardwood and softwood skirtings is differentiated in the specification description.

AA38 is the taking off for walls, starting by measuring overall the plaster and decoration to all rooms. The lengths of walls, as shown on Figures 6.9 and 6.10, were previously calculated (for the skirtings) and the height taken directly from the drawings (2400mm). Note the timesing (multiplying) up of the dimensions for, first, the number of rooms of that size in the dwelling and, second, for the number of dwellings in the block. Signposts again note what dimension applies to which location. A figure "9" at the bottom of the run of dimensions is a check that the correct number of rooms has been measured for each dwelling – a useful overall check that, at least, a whole room has not been forgotten. The NRM2 rules dealing with finishes to walls were explained in the introduction – finishings are covered in section 28, work to walls at 28.1, areas over 600mm wide

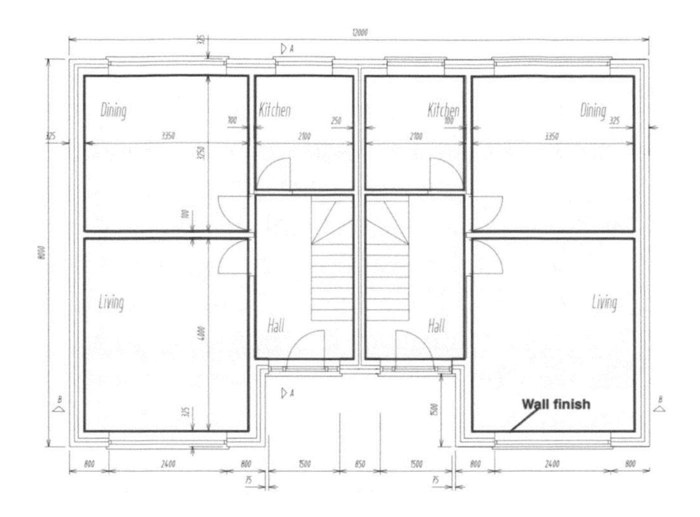

Figure 6.9 Wall finish, ground floor

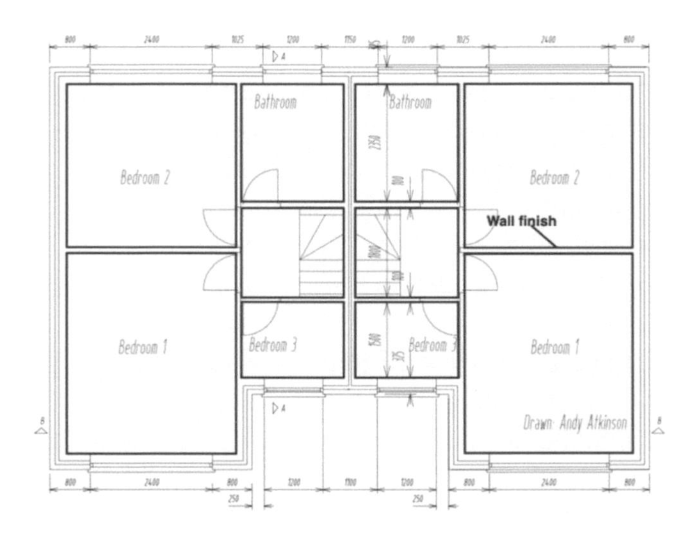

Figure 6.10 Wall finish, first floor

at 28.1.2 and the nature of the base in the general rules to the section. Thereafter, the specification is for standard two-coat work 12.5mm thick using gypsum-based plaster. “Carlite Browning” and “Carlite Finish” are manufacturer’s trade names, and the specification will further reference this. **Render** and **Set** are plasterer’s terms for the rough base and fine finish coats of plaster, respectively. NRM2 rules for decoration are in section 29, general surfaces (including walls) at 29.1, over 300mm girth at 29.1.2, internally at 29.1.2.1 and the nature of the base in the general rules to the section. The specification is for three coats of emulsion paint, and this will be supplemented by manufacturer’s information and national and international standards.

Following the overall measurement of plaster and paint to the walls are adjustments, first for those rooms which are wallpapered or tiled rather than painted. A deduction is made on the *decoration only* for the living room, dining room, bathroom and hall walls. Additions are then made for wallpaper and ceramic tiling. NRM2 deals with wallpaper at 29.9, for areas of more than 1m^2 at 29.9.2 and the nature of the base in the general rules to the section. The specification will refer to details for preparation and hanging paper, but the supply is dealt with by allowing a Prime Cost Sum. The estimator will price supplying and fixing the paper based on this supply price, but the price will be adjusted in the **final account** based on the actual cost of the paper when supplied. NRM2 deals with ceramic tiling in the finishings section as a wall finishing, 28.7; over 600mm wide, 28.7.2; and the nature of the base in the general rules to the section. As with wallpaper, the supply of tiling is expressed as a Prime Cost Sum (£30.00/m^2) to be adjusted in the final account. A further deduction of emulsion paint and addition of tiling is made for a tile "splashback" above kitchen units (Prime Cost Sum of £40.00/m^2). As this is only 450mm high (under 600mm wide), it is measured in linear metres in accordance with NRM2 28.7.1.

AA39 is the taking off for adjustments for external openings, the dimensions for which are taken from the floor plans, sections and elevations. For the last, not all figured dimensions are shown, and some opening heights have been scaled. Dealing first with adjustments to the wall plaster: as a deduction on the earlier full item, it is not necessary to reference it to NRM2 or write the item out in full. Following on from the deduction are additions for reveals (the internal exposed walls at the sides of the windows). The item for this work is identical to that for the main walls, except the work is less than 600mm wide (200mm) and is, therefore, measured in linear, not square, metres. The reveal to the soffit (underside) of the steel lintel is measured separately, as it is to a different base (steel lintel) and to a ceiling (NRM2 28.9.1). The total number of openings per dwelling has again been counted as a check and the number (8) entered at the bottom of the column (note – there is one less room with an opening than the total number of rooms per dwelling, as the landing has no windows!).

To ensure that plaster will adhere to the wall plates, at the top of external walls and loadbearing partitions, a narrow strip of expanded metal lathing is measured. The length of this is the total internal girth of the building (less the party wall) plus the length of loadbearing partitions (multiplied by 2 for each side and 2 again for the two dwellings in the block). No adjustment has been made for the short length of mesh where partitions join external walls. NRM2 deals with metal mesh lathing at 28.31, to walls at 28.31.1, depth at 28.31.1.1, method of fixing and nature of background at 28.31.1.1.1+2.

The arris (external angle) beads between window reveals and main plaster are measured next, with horizontal lengths over all openings taken first, followed by vertical lengths to the jambs. Although the nature of each background is required by NRM2, the background has all been taken as masonry on the basis that the method of fixing (to plaster dabs before full plastering) to the base is the same for both vertical (masonry) and horizontal (masonry/steel) beads. NRM2 deals with arris beads at 28.28.1, the requirement for a "dimensioned description" being satisfied by referring to a separate specification.

Adjustments to decorations for external wall openings have been taken separately from adjustments to finishings. First, emulsion-painted walls are adjusted by deducting window areas in the kitchen and bedrooms. The areas are reduced (using a "LESS" item) to allow for painting the reveals (NRM2 section 29, decorations, *does not* require work in narrow areas or to reveal soffits to be given separately). On AA40, the same procedure is adopted for the wallpaper in the dining room, living room and hall. Tiling in the bathroom is dealt with similarly, but, as it is a wall finishing, it is measured in accordance with NRM2 section 28. The soffit reveal is categorised at 28.9.1 and the jamb reveals at 28.7.1, both measured in linear metres stating the width. The external angle between window reveal and main tiling has a special "rounded edge" tile, measured in linear metres on AA41 in accordance with NRM2 28.23.1. The same rounded edge tile is measured for the top of the splashback in the kitchen – along the top and down two sides. A special "REX" corner tile has been measured at the junction of horizontal and vertical edges.

Adjustments for internal doors are taken in the right column at AA41, starting with deducting plaster and paint to all rooms. There are seven internal doors in total, two sides to each door and two dwellings in the block. Adjustments are then made for rooms with wallpaper by deducting the paper and adding back emulsion paint previously deducted. Ceramic tiling in the bathroom is dealt with in the same way – by deducting tiling and adding back the emulsion paint previously deducted. Finally, a note is added in the right column of AA42 that no adjustment has been made for decorative finishings behind skirtings – that is, the wall finishings have been measured down to the floor level, leading to a slight over-measurement.

Key points covered in this chapter:

- Outline of the technology of traditional ceiling, floor and wall finishes in the United Kingdom
- Names and nature of some traditional finishes, such as lath and plaster ceilings, render and set walls
- Identifying items for taking off internal finishes
- Measuring openings – the form for deducting quantities
- Boundary items associated with finishes, such as skirtings, coving, reveals and arrises
- Constructing item descriptions for internal finishes using NRM2
- Applying explanations to the example

Note

1 No deduction for ceiling finish for the hatch would be required, as voids of less than 1m^2 are not deducted – NRM2 29.10, notes.

7 Windows, doors and internal fittings

Windows and doors

AA43 to AA46 is the taking off for windows and doors, both external and internal. Although the taking off is straightforward for this element, as with other sections, it is important to understand the technology involved before tackling the measurement. Until recently, the predominant material used for windows and external doors was timber. Although timber is still extensively used, particularly for good-quality work, for much house-building, it has been supplanted by factory-made, glazed and finished PVC.

Much timber joinery, particularly windows, has long been prefabricated, with semi-completed units being delivered to site for installation, glazing, fixing ironmongery and painting. Timber window and door frames may have final assembly on site (for example, by fixing hinges and **sashes**), but the basic unit would be ready assembled. Timber windows and external door frames are traditionally "**built in**" to external walls as these are raised. That is, the unit is set on the wall, and brickwork or masonry is built up around it. Fixing the unit to the wall is by using **frame ties** – strips of metal screwed to the frame and built into the masonry joints. Often the window or door frame would also perform a structural function by supporting the outer skin of cavity walling over the unit and so used substantial sections of timber.[1] As the bricklayer or mason is building in the frame, there is no problem with the unit not fitting the opening; a great advantage over using separate units installed into **blank openings**, as, for the latter, any mistake in measurement would lead to the unit not fitting. With the advent of sealed double-glazed panes of glass, site glazing has been simplified (no longer is most glass cut on site) at the cost of having to perfectly match off-site produced panes to openings. Also, **ironmongery** for volume timber units is usually fixed at works, reducing joinery work on site. Nevertheless, for good-quality work using hardwood joinery, or for historic repair and restoration, much detailed work remains site based.

Although most timber external door frames are now pre-fabricated, hanging external doors into built in frames, fixing ironmongery and painting remains site based. However, internal doors are traditionally hung to **door linings** fixed into blank door openings after the walls are built.[2] Linings may be site cut or supplied as pre-cut **sets**. Using complete **door sets** is also common – the pre-cut lining, **butts**, ironmongery and door are provided as a set for final assembly on site.

The taking off for timber windows, doors, linings and frames follows logic. Where the item is supplied as a unit – for example, a window unit, door frame or internal door set – the item is **enumerated**. Where items are supplied loose for cutting and assembly on site – for example, separate door linings, **door stops**, **architraves** and **window boards** – these are measured in linear metres.

PVC window and door units, designed to reduce site work and maintenance to a minimum, are self finished and manufactured off-site incorporating glass and ironmongery. As relatively delicate, precision-made items, they will not withstand building in but must be carefully installed to blank openings. The method of fixing is by **plugging and screwing** to brick or masonry jambs

DOI: 12.01/9781003253129-8

and, to ensure an exact fit, suppliers will provide matching **templates** to site. These are loosely built in and removed when the units are ready for installation.

In our example, although the overall sizes are known, the detailed type and configuration of PVC units to external walls is yet to be decided. The cost of supply is, therefore, dealt with as a **Defined Provisional Sum**. This sum is an allowance, made in the **Bills of Quantities** and **Contract Sum** for supply, to be adjusted when final decisions on units have been made and the exact cost is known. As the size and overall nature of each unit is known, prices can be obtained for fixing units. The total cost of the installation will be the agreed invoice for supply, set against the Defined Provisional Sum, plus the prices for fixing. AA43 starts, therefore, with the item incorporating the Provisional Sum.

NRM2 deals with Defined Provisional Sums in the glossary of definitions on page 7:

> A sum provided for work that is not completely designed but for which the following information is provided:
>
> - the nature and construction of the work
> - a statement of how and where the work is fixed to
> - the building, and what other work should be fixed
> - a quantity or quantities that indicate the scope and extent of the work, and
> - any specific limitations, etc. identified.

Following on from the Provisional Sum are the items for fixing the units. The specification referenced in the items would include what exactly is to be priced as "fixing" – for example, the exact method of handling, assembling and installing the units. Note the method of entering dimensions. Each dwelling is dealt with separately, and then all units are timesed (multiplied) by 2 to allow for the two dwellings in the block. That way, it is readily clear that there are 8 units per dwelling and 16 units in total.

In our example, window boards are not provided as parts of the window units but are site cut and fixed. Boards typically run past the end of the window reveals by about 75mm (called the "stooling"), and this length is allowed on the timber. NRM2 deals with window boards at 22.5, with the overall cross-section at 22.5.1 and number of labours at 22.5.2. There are two labours on the window board – rounding the front edge and providing a tongue to the rear edge to fit the window unit. The method of fixing, by plugging and screwing, is not at the discretion of the contractor, so it is also specified in the item in accordance with the general rules to the section. Painting the window board is as a general surface at 29.1, less than 300mm girth at 29.1.1 and as internal work at 29.1.1.1.

All internal doors are the same size to suit an 800 × 2000–mm opening with a 32-mm lining – making a net door size of 736 × 1968mm. NRM2 deals with doors at 24.2, requiring a dimensioned description at 24.2.1. The specification and national and international standards will amplify brief details given in the item. In our example, door linings are site cut, fixed and specified to be plugged and screwed to masonry. NRM2 itemises linings at 24.10, requiring a dimensioned cross-section at 24.10.1 and any labours stated at 24.10.1.1. The linings are plain rectangular sections, with no labours worked on the timber, but NRM2 general rules to section 24 require that the required method of fixing (plugging and screwing to masonry) be stated. See Figures 7.1 and 7.2 for details of a typical internal door jamb.

Plain linings require a small section timber (50 × 15mm) door stop to restrain the door when closed. This is measured at the top of the right column on AA44. NRM2 deals with stops at 24.11, with a requirement to state the cross-section at 24.11.1 and any labours (of which there are none) at 24.11.1.1. The length taken for the stops is the same as the lining, although it is actually very slightly shorter, as it runs on the inside of the lining. Also associated with the linings are architraves to both sides. Architraves extend beyond the face of the lining to the extent of the width of the architrave (60mm) minus the lap over the lining (10mm) giving an extra 50mm × 2 greater length at each side of the two top corners. The addition to its length in the waste calculation of "2/2/50" is familiar as a girthing calculation taking the measurement of the architrave to the external length. Architraves have been measured to both sides of all seven doors in the two dwellings but have been adjusted by omitting 2m lengths to jambs where a full-width architrave cannot be fitted in. Examining the floor plans reveals that this occurs at the closing edges of all seven doors in each dwelling plus where two door openings abut each other as shown in Figure 7.2. In place of the full architrave, a small section (20 × 20mm) **quadrant** is measured, which will fit between lining and adjacent wall or lining without obstructing the door. Note that NRM2 deals

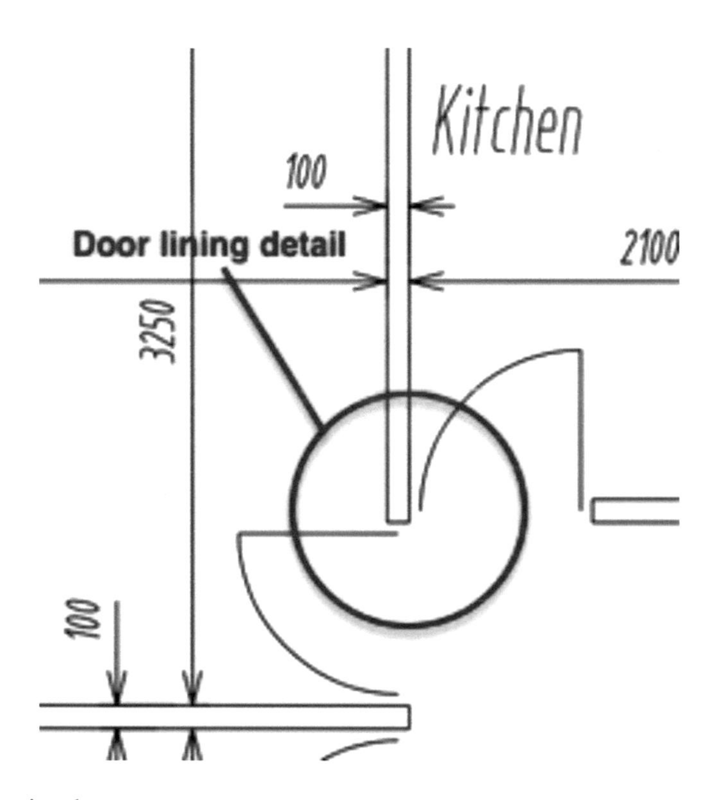

Figure 7.1 Internal door jamb

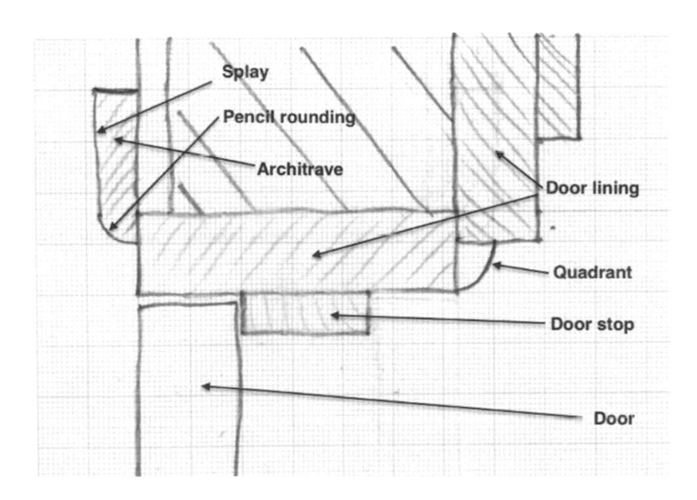

Figure 7.2 Door lining detail

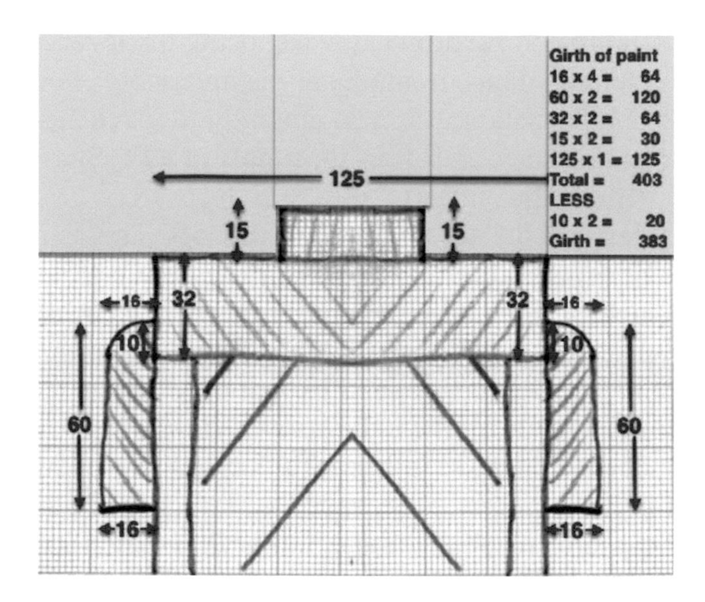

Figure 7.3 Paint girth of door lining

with architraves at 22.2, with a requirement to give a dimensioned cross-section at 22.2.1 and the number and type of labours worked on the item (described in the item as "**splayed and rounded**" and illustrated in Figure 7.2) at 22.2.2. The quadrant is measured at 22.3 – "cover fillets, stops, trims, beads, nosings, etc." – giving the cross-section size at 22.3.1 and the type of labour (as a "quadrant") at 22.3.2. Note that it is unusual for architectural details as shown in Figures 7.2, 7.3 to be provided, and the taker off will visualise the detail based on the specification and personal knowledge of construction technology.

The waste calculation immediately following the item for the quadrant is for the area of paint on doors. Assuming that the face and one edge of each door is painted on each side, this gives an exact area of 2006 × 774mm. The taker off has rounded this to the net area of the blank opening – 2000 × 800mm. NRM2 deals with paint to doors as a general surface at 29.1, over 300mm girth at 29.1.2 and internal work at 29.1.2.1, with the nature of the base (plywood) given in accordance with the general rules to the section. The area measured is both sides of all seven doors to both dwellings. To calculate the area of painting for the lining (and to determine whether it is measured in m or m^2 in accordance with NRM2), the next waste calculation girths up the widths of all the painted elements of the lining – the lining itself (125mm), the edges of the lining (2 × 32mm) two sets of architraves (2 × 60mm), four edges to the architraves (4 × 16mm), the two sides of the door stop (2 × 15mm) less the overlap of the architrave over the lining (minus 2 × 10mm). This amounts to a girth from architrave to architrave of 383mm, which is more than 300mm – meaning that the paint to the lining is measured in m^2 in accordance with NRM2 29.1.2.1. The area of paint taken for the linings is, therefore, the length of the door head and two jambs multiplied by the girth of 383mm multiplied by the number of linings in the dwelling and of dwellings in the block. A note above the item advises that no adjustment has been made for situations where a quadrant has been measured in instead of an architrave. So, in Figure 7.2, the girth of painting is still taken as 383mm for each jamb even though it is clearly slightly less where two jambs are next to each other.

Ironmongery appears in three sections of NRM2–22, general joinery; 23, windows etc.; and 24, doors etc. NRM2 24.16 deals with ironmongery to doors, with a requirement to state the type of item, method of fixing and nature of base at 24.16.1+2 and the type and quality of material and fixings etc. at 24.16.1.1–3. For standard internal doors to dwellings, ironmongery consists of three items – the **butts**, **latch** and **furniture**, the last being the levers and knobs visible on the face of the door. As manufactured items, ironmongery is invariably enumerated and specified by reference to manufacturer's information. It is also common to leave the detailed specification of ironmongery to the post-contract stage and allow for supply as a PC Sum (similar to the wallpaper and wall tiling measured with the finishings).

The next set of items relates to the access hatch to the loft. This is not shown on any drawing but is mentioned in the specification. Where traditional built-up roof construction is used, the hatch will be positioned on the landing near a loadbearing partition to reduce the **bending load** on the joists concerned. It is formed by trimming one ceiling joist, normally giving a blank opening size of 750 × 750mm (designed to trim between joists at 400mm centres with one joist removed). The hatch itself can be a purpose-made PVC unit enumerated in the same way that the windows are measured or, as in our case, formed from a cut down door panel, with plain lining to the opening and standard architrave as trim. NRM2 deals with hatches at 24.6, with a requirement to give a dimensioned description or diagram at 24.6.1. The hatch lining is measured in the same way as door linings in accordance with NRM2 24.10.1.1 and the architrave with NRM2 22.2.1. Painting to the hatch is the same as to internal doors but is in a small isolated area of less than 1m^2. NRM2 requires that such small areas be enumerated at NRM2 29.1.3.1. Painting the lining and architrave is the same as for the door linings, but the girth of the work is less than 300mm, so the paint is measured in linear metres as NRM2 rule 29.1.1.1. There is no deduction for the finish or paint to the *ceiling* for the hatch (in accordance with NRM2 28 General rules note 3 and 29.10, note 1), as the area involved is less than 1m^2. Trimming the opening requires two trimmers 750mm long and six joist hangars measured in the same way as in the Roof Construction section (Figure 7.4). The short length of joist cut out has not been deducted (and a note has been provided in the taking off to this effect), as the amount of timber involved is minimal.

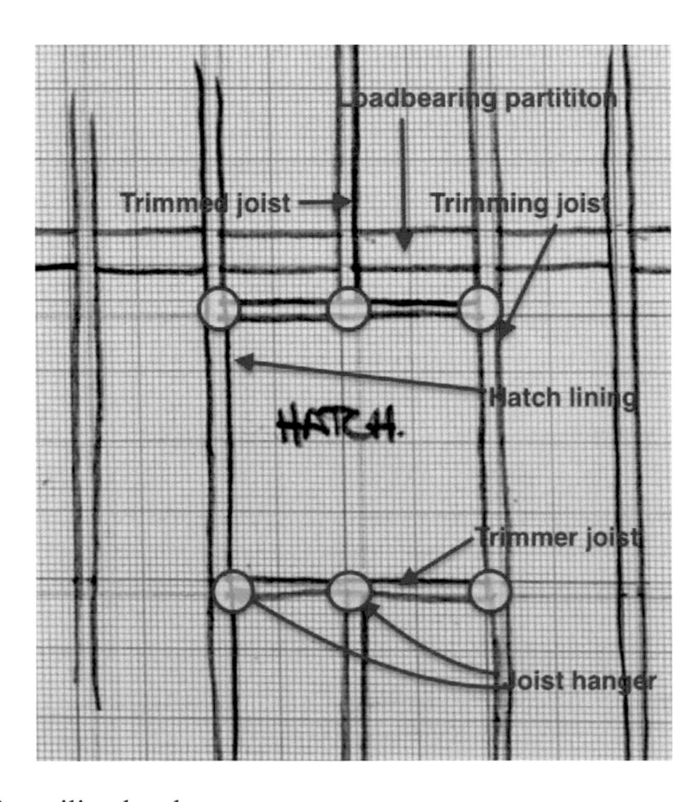

Figure 7.4 Trimming for ceiling hatch

Internal fittings

AA47–AA49 are the taking off for internal fittings. Very little is shown on the drawings or mentioned in the specification, but there would normally be fittings in the kitchen and bathroom, possibly fitted wardrobes in bedrooms, drying cupboards, larders and so on. The staircase and associated balustrading are, however, shown in outline on the drawings and referenced in the specification. The staircase would normally be supplied as a preconstructed joinery item from a specialist joiner, even if designed in detail by the architect. In practice, even where the architect is designing the unit, he or she will not provide detail but will provide outline designs for the visible structure, leaving the detail to the joinery shop.

Whether design is in detail, outline or left to the joiner, it is common to defer detail design to the post-contract stage and allow a Provisional Sum for supply and fitting of the unit. Clearly, the size and configuration of the staircase is apparent from the drawings. This gives sufficient information to classify the sum as a Defined Provisional Sum, the requirements for which are mentioned at the start of this section in relation to the supply of PVC windows. This means that the tendering contractors can allow for the provisional sum plus price for preliminaries associated with installing the staircase. This would not include specific items of "**builder's work**" related to the staircase, for example, in painting the item, providing plasterboard to the underside of the stairs and working finishings and skirtings around the unit, but would include providing general support, attendance, scaffolding and so on. Where identified sufficiently from the drawings, builder's work can be measured as firm or Provisional Quantities.

AA47 starts by itemising the provisional sum and follows with the builder's work in connection. First, the stairwell areas for finish and decorations to the ceilings are deducted (note that it is not necessary to deduct the first-floor boarding, as this deduction was made when the flooring was measured). At the same time, areas for the same items are deducted for work to first-floor ceilings, which are more than 3.5m above finished or structural floor levels. This occurs in the "drop" to the stairwell. NRM2 requires that work over 3.5m high be measured separately, as this work is more expensive to carry out, and the following items add back the same quantities, with amended descriptions.

The first item at the top of AA48 deducts the cove to the ceiling at the side and end of the staircase opening, and the following items are for the finishings and decoration to the underside of the stairs. As no details of these items have been provided, the taker off has assumed the construction as basically the same as for other ceiling areas (plasterboard, skim coat plaster and emulsion paint), made an approximate estimate of the areas and lengths involved and marked the items as a "**Provisional Quantity**". Marking items as provisional indicates to the estimator that the work is subject to re-measurement in the final account, using the rates provided against the items at the time of the estimate. NRM2 explains the use of provisional quantities on page 47:

> Where work can be described and given in items in accordance with the tables, but the quantity of work cannot be accurately determined, an estimate of the quantity should be given and identified as a 'provisional quantity'.

Two items are measured for the finishes – the general soffit area of the staircase in m^2 (NRM2 28.9.2) and treatment to the underside of the treads and risers to the **winders** in linear metres (NRM2 28.9.1). Decoration to the soffit is the same as for general ceilings, but, as the item is marked as a provisional quantity, it will be itemised separately.

A vertical board to the edge of the stairwell is the next item measured in accordance with NRM2 22.3.1+2. This is followed by an item for painting all the general areas of the staircase sides, steps and edge of the stairwell – all marked as provisional and subject to re-measurement in the final account. On AA49 the taker off has assumed that the balustrade to the staircase will be of separate balusters, 800mm long at 100mm centres, and has taken a provisional quantity for painting these. Finally, for other unforeseen work in association with the staircase, the taker off has included an Undefined Provisional Sum in accordance with the definitions on page 12 of NRM2.

Finally, no details had been provided at the time of the take-off for kitchen, bathroom and other fittings, so a To Take note was entered on the right column of AA49. Prior to tender, to ensure amounts were allowed in the Bills of Quantities for these items, three provisional sums were included:

- Kitchen fittings – a Defined Provisional Sum on the basis that the nature, extent and configuration of standard kitchen fittings can be determined from the layout of the kitchen and the nature of the project.
- Bathroom fittings – a Defined Provisional Sum, also based on the nature of work being reasonably apparent.
- Builder's work in connection with the kitchen and bathroom fittings – an Undefined Provisional Sum, as this is an allowance for unforeseen work associated with the installations.

NRM2 Explains an Undefined Provisional Sum as "A sum provided for work that is not completely designed, but for which the information required for a defined provisional sum cannot be provided". The significance of the sum being "undefined" is that, should any be required, the contractor will be able to claim additional on-costs (preliminary items) associated with an expenditure against the sum.

Key points covered in this chapter:

- Outline of the technology of traditional and modern windows and doors
- Names of components, including frames, linings, door stops and architraves
- The use of provisional sums for the supply of units and describing "Fix" items
- Identifying items for taking off windows, doors and associated trim
- Constructing item descriptions for windows, doors and fittings using NRM2
- Using Defined and Undefined provisional sums for internal fittings
- The use of provisional quantities
- Applying explanations to the example

Notes

1 Structural window and door frames can prove problematic when they are replaced with PVC, as PVC may not be strong enough to support masonry.

2 Timber door **frames** are used in external walls, are designed for building-in and are **rebated** to resist the weather and take the door. The **head** and **jambs** are jointed using **tenons** and are of substantial timber. Timber linings are set in internal walls, are designed for "lining" a blank pre-formed opening and are not subject to the external environment. Head and jambs are jointed with simple **tongued joints** and are of lighter timber.

8 Bills of quantities

The measured works section of the bills of quantities (BQs) for our project is in Appendix 2. Although we are mainly concerned with mastering taking off, understanding how the output is presented to estimators for pricing is important context. As mentioned in the introduction, modern bill production uses computers, and the arithmetical aspects of BQ production are easily dealt with. Provided the items and variables, based on the classifications of NRM2, are accurately entered, the software will collate the output into **standard bill order** or any other order programmed into the coding. Alternative orders include elemental and operational, the former breaking down the quantities into broad functional elements for approximate estimating and cost planning purposes. The latter breaks the quantities into operations based on the construction sequence, useful for construction programming and resource management. We are only concerned here with standard bills of quantities order, where items are sorted for ease of pricing by estimators.

The standard order loosely follows the construction sequence and starts by breaking down work into **work sections**, some of which are based on the traditional trades of the construction industry. NRM2 has been developed on this basis, so most modern bills in the United Kingdom adopt the NRM2 categories and coding system. The reason for a work section breakdown is that the sections are often carried out by specialists, and, in pricing, the relevant BQ section will be passed to and priced by the specialist, with the overall contractor's estimator collating and managing the output. Within the work section, categorisation will be by grouping similar specifications – so, for example, all the facing brickwork with the same specification will be grouped together. For this categorisation, following the *order* of sub-categories of NRM2 will be less rigid, but the items within the category will still follow NRM2 requirements. Within those items with the same specification, items are ordered to facilitate easy and progressive pricing by estimators. So, where relevant, cubic (m^3) items are billed first, then square (m^2), then linear (m) and finally enumerated (No). An example showing the progressive nature of pricing is traditional painting, where the estimator will price broad general surfaces in m^2 first; from that price will be derived a price for narrow lengths of less than 300mm girth in metres, and from that price will be derived a price for an enumerated small area of less than $1m^2$.

In standard BQ order, items will only appear once, making for less repetitious pricing (the exception is substructures, which are kept completely separate and may repeat items from other sections). Showing the total quantity involved in any item is also valuable in pricing – large quantities may command bulk discounts from suppliers or dictate types and sizes of plant. However, *measurement* of an item may take place in several places in the taking off and may involve deductions as well as additions. This gives rise to the need to "gather up" quantities into one item, sum the additions and deductions and calculate the net quantity. Computer quantities can easily handle these operations, but, for our examples and for illustration purposes, this has been carried out manually on a "working up"[1] copy of the taking off (Figure 8.1). On this copy, as items have been transferred to the bills, they have been shaded yellow. Prior to the

DOI: 12.01/9781003253129-9

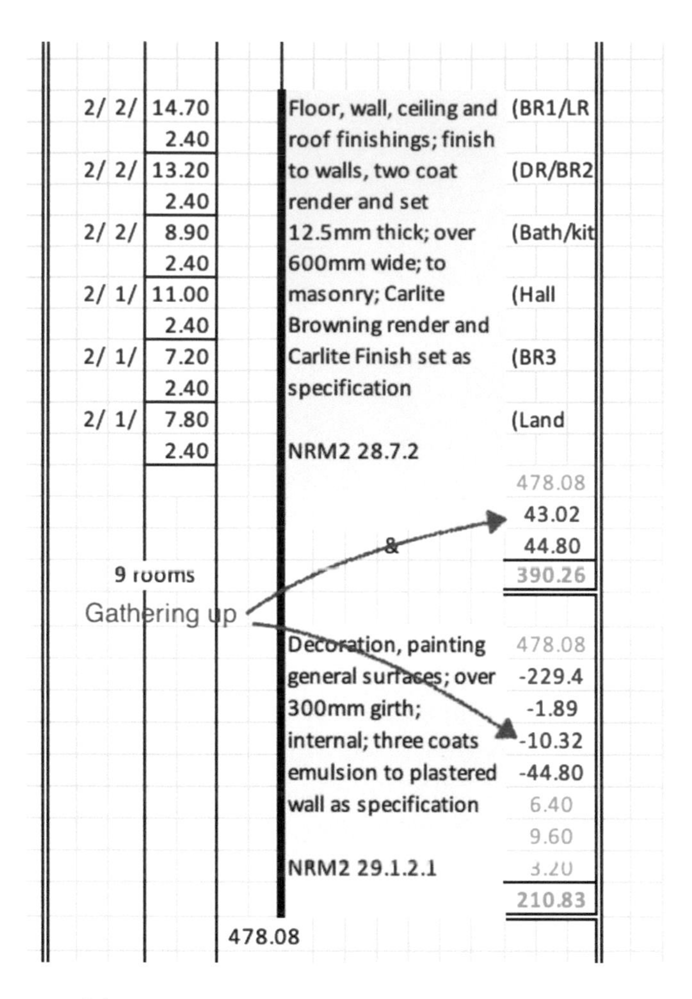

Figure 8.1 Gathering up quantities

transfer, identical items appearing elsewhere in the taking off have been gathered up and the net quantity calculated.

Not all price information relevant to a project is contained in measured works, and bills of quantities include further general and specific information:

- Preliminaries
- General specification
- Provisional sums

Preliminaries are those items having a cost significance but related to the project as a whole or for which pricing in a measured work item is not possible. Examples include site accommodation, scaffolding, site hoardings, managerial personnel and insurance. The taker off will also bring to the estimator's attention particular cost significant features of the project – restrictions on access, hours of working, methods of financing and so on. NRM2 gives detailed guidance and requirements for preliminaries in Work Section 1. Examples include the requirement to give a brief description of the site (NRM2 1.A3.1) and identify general constraints on executing the works (NRM2 A5.1) – for example, information on access, parking, storage and deliveries. Opportunities must be given to price site specific items such as temporary accommodation (NRM2 1.B5.1) and temporary electrical supplies (NRM2.1.B6.3).

A general specification will give details to back up measured work items. It is good practice in drafting bills to make the measured items as brief as possible, provided essential specification and dimensions are included. All other background information will be contained in general specification clauses, which in turn may refer to reference documentation not included in the bills, such as an architect's specification or national or international standards. Referring to external documents, in contractual terms known as "incorporation by reference", is an effective way of ensuring that all important aspects of an item are included but without making the bills of quantities too cumbersome. NRM2 contains a requirement that the preliminaries refer to any specification either included or referred to in the bills (NRM2.1.A2.1).

Provisional Sums have been encountered in our example in relation to windows, external doors, the staircase, and kitchen and bathroom fittings. They are an allowance for work that may be foreseen in some detail but not fully designed or for work that may be needed, but the extent and nature of the work is uncertain. Provisional sums for the former are called Defined Provisional Sums and the latter Undefined Provisional Sums. An example of a Defined Provisional Sum is that for the supply of windows and external doors in our example. The nature and size of the units is known sufficiently for the items to be handled and installed on site, but the exact type of unit is yet to be decided or priced. An example of an Undefined Provisional Sum is that for the builder's work to the installation of the specialist staircase. Although the nature of the staircase is known with some detail, the extent and nature of work associated with the installation is very uncertain. The significance of the distinction between defined and undefined provisional sums is that, for the former, it will be assumed that the contractor has priced for preliminaries associated with the item. For the latter, this will not be assumed and, should costs arise in respect of extra preliminaries, then they will be added.

Modern bills of quantities may also contain other sections, and NRM2 gives guidance on the overall composition of bills of quantities at 2.5 on page 18:

> BQs usually comprise the following sections:
>
> - form of tender (including certificate of bona fide tender)
> - summary
> - preliminaries
> - measured works
> - non-measurable works
> - provisional sums
> - contractor-designed works
> - risks
> - credits (for materials arising from the works)
> - dayworks (provisional) and
> - annexes

Although many of these sections are not relevant to the simple bills of quantities for our project, one section that has become quite common is **Contractor Designed Works**. Some elements within otherwise traditional bills of quantities projects are now frequently designed by the tendering contractors rather than being specified on drawings, in a specification or as a BQ item. These elements could include the frame for a building or electrical, heating, plumbing and air conditioning systems. The BQ will make provision for the element by itemising "Employer's Requirements" – usually in the form of performance specifications and/or outline drawings – but not in the form of measured items as required by NRM2. Some Employer's Requirements may, however, include the requirement that a chosen contractor produce quantities based on the proposed design and in accordance with NRM2.

This book concentrates on taking off building work and not on other elements of documentation, such as drafting preliminaries or Employer's Requirements. It is important to understand the place of these activities in preparing bills of quantities, but carrying them out can be learned once the basic skill of measurement has been mastered.

Key points covered in this chapter:

- Order of items in bills of quantities
- An outline of the process of working up applied to our example
- The overall contents of bills of quantities
- The role of preliminaries
- An outline for constructing items for preliminaries using NRM2
- The role of general specifications and how they are incorporated in bills of quantities
- Provisional sums for defined and undefined work
- Incorporating contractor-designed work in bills of quantities

Note

1 "Working up" is the process of converting the taking off to bills of quantities. Prior to computerisation, working up was a considerable activity involving squaring the dimensions, re-writing taking-off items in bill format, gathering and reducing quantities to one per item.

Appendices

Appendix 1

Taking off

Semi-detached houses 1		AA1		Substructure 1
	Job No 204 NRM2			
	Taker-off - Andrew R Atkinson			
	Section - Substructure			
	Date - 12-02-2022			North - South
	Drawings Q1/A4/2667/ 1a			
	3a		8000	325
	6f			4000
	7c			100
	2c			3250
				325
	T.O. List			
				8000
	Vegetable soil			
	Dispose of vegetable soil			
	Excavate trenches			
	Disposal of excavated material			
	Earthwork support			
	Concrete			
	Blocks and bricks			
	Damp-proof course			
	Earth backfilling			
	Hardcore backfilling			
	Hardcore bed			
	Concrete bed			
	Tamping concrete			
	Damp proof membrane			
	Dimension check			
	East - West			
12000	325 800			
	3350 2400			
	100 800			
	2100 75			
	250 1500			
	2100 850			
	100 1500			
	3350 75			
	325 800			
	2400			
	12000 800			
	12000			

Semi-detached houses 2			AA2			Substructure 2
			Vegetable soil			
				12.30		Excavating and filling;
			12000 x 8000	8.30		site preparation; remove
				DDT		topsoil 150mm deep.
		625		3.70		
	minus	325		1.50		
						NRM2 5.5.2
2 /	0.5 /	300	300 300			
			12300 8300			
			Deduct re-entrant at entrance			
			75			
			1500			
			850			
			1500	12.30		Excavating and filling;
			75	8.30		filling obtained from
			4000	DDT		excavated material;
				3.70		150mm thick; topsoil,
	minus			1.50		20m distant, rotovated
						and selected.
2 /	.5 /	300	300			
						NRM2 5.11.1.1.1
			3700 x 1500			

Semi-detached houses 3 | **AA3** | **Substructure 3**

Foundations

Girth of walls

Across building]		12300
Down building]		8300
		20600
		2
		41200
Re-entrant] 2 / 1500		3000
Edge of topsoil]		44200
4 / 2 / 0.5 / 625		2500
Centre-line of outside wall]		41700
Add for party wall		
Down building]	8300	
Less re-entrant]	1500	
	6800	
Found] 2 / 625	1250	
Net length of party wall]	5550	5550
Total length of 625 trench]		**47250**

Partition trench

		3350	
		100	
		2100	
Scaled]		900	
Total partition]		6450	
625			
325			
0.5 / 300	150		(a)
625			
250			
0.5 / 375	187.5		(b)
	337.5	337.5	
		6112.5	

(a) = Spread of outside wall foundation

(b) = Spread of party wall foundation

Depth

	1000
Vegetable soil]	150
	850

Semi-detached houses 4			AA4				Substructure 4	
	47.25	Excavating and filling;						
	0.63	excavation starting						
	0.85	150mm below						
2 /	6.11	existing ground level;						
	0.40	foundation	[Partitions					
	0.85	excavation not						
		exceeding 2.0m deep						
		NRM2 5.6.2.1						
					47.25	Insitu concrete works;		
					0.63	substructure; plain		
					0.40	insitu concrete;		
		&		2 /	6.11	horizontal work;		
					0.40	over300mm thick; in	[Ptns	
					0.40	structures; poured on		
		Excavating and filling;				or against eaerth or		
		disposal; excavated				unblinded hardcore;		
		material off-site.				mix 15/N/mm2 -		
						40mm aggregate.		
		NRM2 5.9.2						
						NRM2 11.2.2.2.1		

Earthwork support.

2/	47.25	Excavating and filling;	[Main tr
	0.85	support to faces of	
	44.20	excavation;	[Edge TS
	0.15	maximum depth less	
2/ 2/	6.11	than or equal to	[Ptns
	0.85	1.00m; trench	
	DDT	foundations; distance	
2/	0.63	between opposing	[End PW
	0.85	faces less than or	
2/ 2/	0.40	equal to 2.00m.	[End ptn
	0.85		
		NRM2 5.8.1.1.3	

Semi-detached houses 5				AA5				Substructure 5	
			Brickwork						
			Girth of walls					**Depth**	
Across building]				12000					
Down building]				8000				**Depth to ground level]**	1000
				20000				**Ground to DPC]**	150
			x	2					150
				40000					
Re-entrant]			2 / 1500	3000				**Less foundation depth]**	400
Outside face]				43000					
	4 /	2 /	0.5 / 100	400					-250
C/L Outside skin]				42600					
				400					
				42200					
	4 /	2 /	0.5 / 75	300					
C/L Cavity]				41900					
				300					
				41600					
	4 /	2 /	0.5 / 150	600					
C/L Inner skin]				41000					
				600					
Inner face of wall]				40400					
			Party wall						
Depth of building]			8000						
Less re-entrant]			1500						
			6500						
	2 /	325	650 [Less outside wall						
			5850 [Net length						
	2 /	150	300 [Extra @ inner skin						
			6150 (Party wall cavity						
			Partition						
Brought forward from foundations				6450					

Semi-detached houses 6			AA6	Substructure 6
	42.60 0.75	Masonry; brick/block walling; walls; 100mm thick; blockwork; skins of hollow walls; stretcher bond; dense concrete (10.5N/mm2) blocks in cement mortar (1:4). NRM2 14.1.2.1.1	[Outer	
2 /	5.85 0.75		[Party	
	41.00 0.75	150mm Thick ditto.		
2 /	6.45 0.75	Masonry; brick/block walling; walls; 100mm thick; blockwork; stretcher bond; dense concrete (10.5N/mm2) blocks in cement mortar (1:4). NRM2 14.1.2.*.1		

Semi-detached houses 7			AA7	Substructure 7
41.90		Masonry; brick/block walling; forming cavities 75mm wide; stainless steel wall ties spaced 600mm horizontally and vertically.		
0.75		NRM2 14.14.1.1		
6.15		Ditto 50mm wide [Party wl		
0.75				
		Cavity filling		
		Brickwork depth 750		
		Less DPC-top] 150		
		600		
41.90		Plain insitu concrete; substructure; vertical work; less than or equal to 300mm thick; in structures; concrete 21N/mm2 - 10mm aggregate.		
0.08		NRM2 11.5.1.1		
0.60				
6.15				
0.05				
0.60				

Semi-detached houses 8			AA8			Substructure 8
		Facing bricks				
		4 Courses of facings x 75mm = 300		42.60		Masonry; brick/block walling; damp-proof courses less than or equal to 300mm wide; single layer vinyl as specification; horizontal.
			2 /	5.85		
				41.00		
			2 /	6.45		NRM2 14.16.1.3
	42.60	Masonry; brick/block walling; walls half brick thick; brickwork; skins of hollow walls; facework one side; Butterly Brown Brindle bricks manufactured by the Butterly Brick Company, bedded, jointed and pointed in cement mortar (1:4) in stretcher bond, pointed with a neat weatherstruck joint as the work proceeds.				
	0.30	NRM2 14.1.1.1.1/2				
		&				
		DDT				
		100mm Block skin of hollow wall in cement mortar as before.				

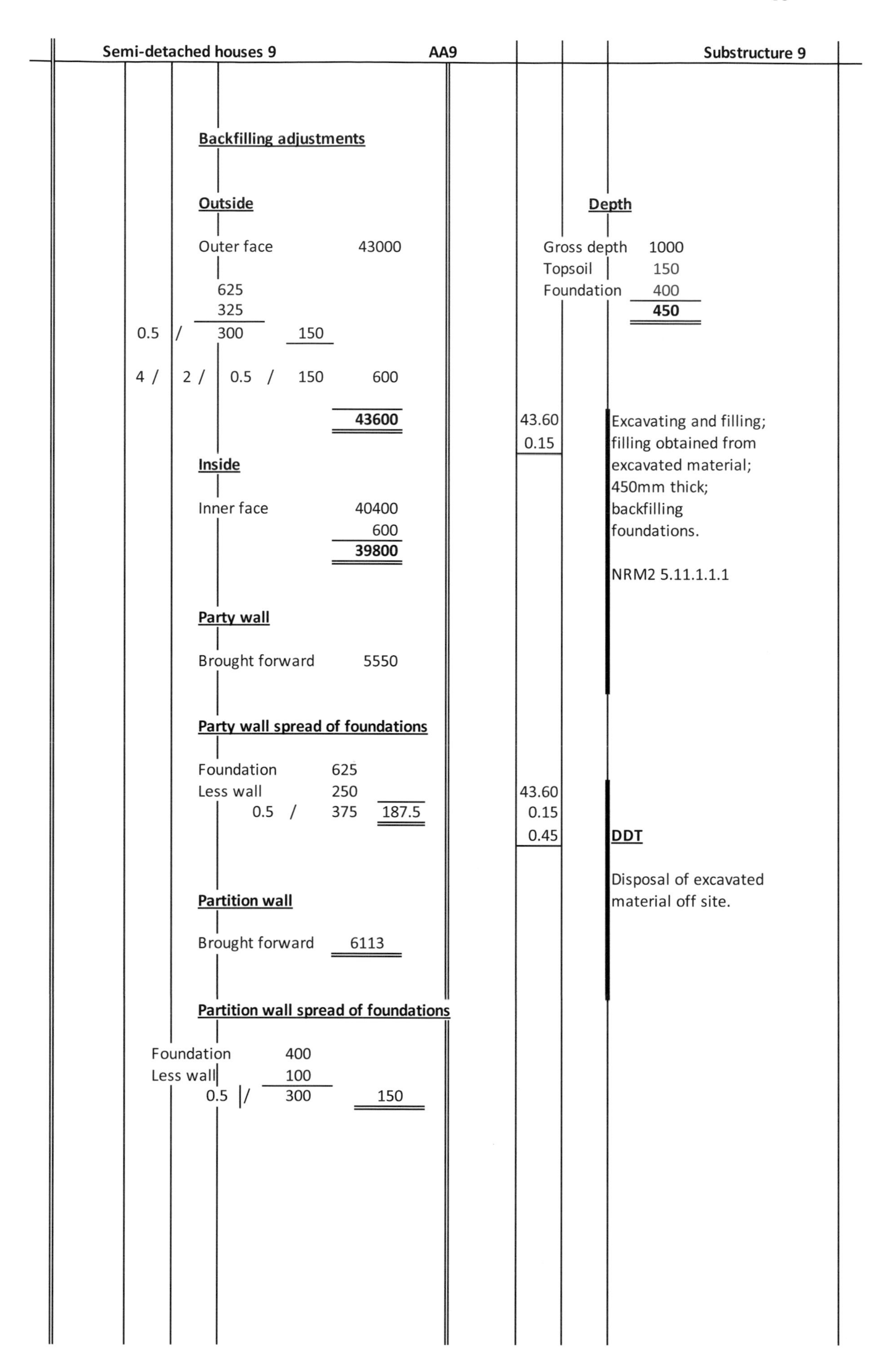
Semi-detached houses 9
AA9
Substructure 9
Backfilling adjustments
Outside
Outer face 43000
625
325
0.5 / 300 150
4 / 2 / 0.5 / 150 600
43600
Inside
Inner face 40400
600
39800
Party wall
Brought forward 5550
Party wall spread of foundations
Foundation 625
Less wall 250
0.5 / 375 187.5
Partition wall
Brought forward 6113
Partition wall spread of foundations
Foundation 400
Less wall 100
0.5 / 300 150
Depth
Gross depth 1000
Topsoil 150
Foundation 400
450
43.60
0.15
Excavating and filling; filling obtained from excavated material; 450mm thick; backfilling foundations.
NRM2 5.11.1.1.1
43.60
0.15
0.45
DDT
Disposal of excavated material off site.

Semi-detached houses 10			AA10	Substructure 10
	39.80	Excavation and filling; imported filling; beds and voids over 50mm thick but not exceeding 500mm thick, 450mm thick; level; backfilling to foundations; hardcore as specification compacted in layers.	[main wl	
	0.15			
	0.45			
2/	5.55		[PW	
	0.19			
	0.45			
2/ 2/	6.11		{ptns	
	0.15			
	0.45			
	DDT			
2/	0.25		[end PW	
	0.15	NRM2 5.12.2.1.1		
	0.45			
2/ 1/	0.10		[end ptn	
	0.15			
	0.45			
2/ 1/	0.10		[end ptn adj PW	
	0.19			
	0.45			

NB No adjustment for topsoil backfilling

Semi-detached houses 11 **AA11** **Substructure 11**

Over-site

		12000 x	8000
Less ext wall	2 / 325	650	650
		11350	7350

Re-entrant

		4000	
Plus ext wall	2 / 325	650	
		4650 x	1500

Timesing	Dimensions	Description	Note
	11.35	Imported filling; beds	
	7.35	and voids over 50mm	
	0.15	thick but not	
	DDT	exceeding 500mm	
	4.65	thick, 150mm thick;	[Re-entr
	1.50	level; over-site bed;	
	0.15	hardcore; clean, hard,	
	5.85	broken brick to pass a	[PW
	0.25	50mm screen as	
	0.15	specification.	
2 /	6.45		[ptns
	0.10	NRM2 5.12.2.1.1	
	0.15		

Timesing	Dimensions	Description	Note
	11.35	Plain insitu concrete;	
	7.35	substructure;	
	0.15	horizontal work less	
	DDT	than or equal to	
	4.65	300mm thick; in	[Re-entr
	1.50	structures; 21N/mm2	
	0.15	- 20mm aggregate.	
	5.85		[PW
	0.25	NRM2 11.2.1.2	
	0.15		
2 /	6.45		[ptns
	0.10		
	0.15		

		Semi-detached houses 12	AA12
	11.35	Plain insitu concrete;	
	7.35	tamping by	
	DDT	mechanical means;	
	4.65	top surface.	[Re-entr
	1.50		
	5.85	NRM2 11.12.1	[PW
	0.25		
2 /	6.45		[ptns
	0.10		
	11.35	Excavating and filling;	
	7.35	damp-proof	
	40.40	membrane; over	
	0.15	500mm wide, 2.7	
2 /	5.85	micron; horizontal	
	0.15	polythene lapped	
2 / 2 /	6.45	300mm at joints on	
	0.15	blinded hardcore.	
	DDT		
	4.65	NRM2 5.16.2.1	
	1.50		
	5.85		
	0.25		
2 /	6.45		
	0.10		

Substructure 12

Semi-detached houses 13 **AA13**

External/internal walls 1

NRM2

Section - External/internal walling/partitions
Date - 12-02-2022

Drawings Q1/A4/2667/ 1a
2c
6f
7c
8
9
10

T.O. List

Facing bricks
Cavity and insulation
Inner block skin
Party wall
Party wall cavity
Loadbearing partitions
Non/loadbearing partitions
Adjustments for extl opening
DDT Facings
DDT Cavity and insulation
DDT Inner block skin
Cavity closer
Lintels
Adjustments for intl opening
DDT Blocks
Lintels

Dimension check (upper floor)

East - West

12000	800	800
	2400	2400
	1025	800
	1200	250
	1150	1200
	1200	1100
	1025	1200
	2400	250
	800	800
	12000	2400
		800
		12000

North-South

8000	1500
	325
	1500
	100
	1800
	100
	2350
	325
	8000

Semi-detached houses 14				AA14			External/internal walls 2
			External walls				
			Girth of walls				
Across building]				12000			
Down building]				8000			
				20000			
			x	2			**Heights**
				40000			
Re-entrant]		2 /	1500	3000			Outer skin/cavity
Outside face]				43000			
	4 /	2 /	0.5 / 102	408			2400
							250
C/L Outside skin]				**42592**			900
							1200
Outside face				43000			**4750**
Outside skin			102				
Half cavity	0.5	/ 73	36.5				
Dist moved	4 /	2 /	138.5	1108			Inner skin
C/L Cavity]				**41892**			2400
							250
Outside face				43000			2400
Outside skin			102				5050
Cavity			73			Plate	50
[Half	0.5 /	150	75				**5000**
Inner sk]							
Dist moved	4 /	2 /	250	2000			
C/L Inner skin				**41000**			
Outside face				43000			
Dist moved	4 /	2 /	325	2600			
Inner face				40400			

			Semi-detached houses 15	AA15			External/internal walls 3
	42.59 4.75		Masonry; brick/block walling; walls half brick thick; brickwork; skins of hollow walls; facework one side; Butterly Brown Brindle bricks manufactured by the Butterly Brick Company, bedded, jointed and pointed in gauged mortar (1:1:6) in stretcher bond, pointed with a neat weatherstruck joint as the work proceeds. NRM2 14.1.1.1.1/2	[Outer	42.59 4.75		Masonry; brick/block walling; forming cavities 73mm wide; stainless steel wall ties spaced 600mm horizontally and vertically. NRM2 14.14.1.1 & Masonry; brick/block walling; cavity insulation; 73mm glass-fibre resin treated wall bats, fixed between wall ties. NRM2 14.15.1.1
							5000 4750 250 Ins above cavity
					42.59 0.25		ADD 73mm Cavity insulation, as before

			Semi-detached houses 16	AA16			External/internal walls 4	
					2/	5.85	Masonry; brick/block walling; walls; 100mm thick; blockwork; skins of hollow walls; stretcher bond; dense concrete blocks (10.5N/mm^2) in gauged mortar (1:1:6). NRM2 14.1.2.1.1	
	41.00		Masonry; brick/block walling; walls; 150mm thick; blockwork; skins of hollow walls; stretcher bond; insulating concrete blocks in gauged mortar (1:1:6). NRM2 14.1.2.1.1			5.00		
	5.00				2/0.5/	6.45		[in roof
						2.80		
					2/	6.45		[in roof
						0.05		[no plate

Party wall

BR3]	1500	
	100	
Landing]	1800	
	100	
Bath]	2350	
	5850	
2 / 150	300	(extra @ inner sk
Party wall cavity	**6150**	

In roofspace

	5850	
2 / 300	600	[at eaves
	6450	

Semi-detached houses 17			AA17		External/internal walls 5
	6.15		Masonry; brick/block walling; forming cavities 50mm wide; stainless steel wall ties spaced 600mm horizontally and vertically.	[Party wl	
	5.00				
0.5/	6.45			[in roof	
	2.80				
	6.45			[no plate	
	0.05				
			NRM2 14.14.1.1		

Semi-detached houses 18			AA18			External/internal walls 6
Partitions						
Loadbearing - Ground floor to first floor						
Living/dining room		3350				
Partition		100				
Kitchen/hall		2100				**Loadbearing**
First floor						
Staircase	1800			2/	6.45	Masonry; brick/block walling; walls; 100mm thick; blockwork; stretcher bond; dense concrete blocks (10.5N/mm2) in gauged mortar (1:1:6).
Partition	100				5.00	
Bedroom 3	1500					
Re-entrant	1500					
	4900					
	4000					
	900	900				
		6450				NRM2 14.1.2.*.1
Non-loadbearing - Ground floor						
	3250					
	900					
Kitchen-dining	2350	2350				
	4000					
	1500					
Living-hall	2500	2500				
Total		**4850**				
Non-loadbearing - First floor						
Bath - bedroom 2		2350				
Bedroom 3 - landing		2100				
Bedroom 1/3 - landing		2500				
		6950				

			Semi-detached houses 19	AA19			External/internal walls 7
			Non-loadbearing				
2/	4.85		Masonry; brick/block	[GF			
	2.40		walling; walls;				
2/	6.95		100mm thick;	[FF			
	2.40		blockwork; stretcher bond; lightweight concrete blocks in gauged mortar (1:1:6).				
			NRM2 14.1.2.*.1				

Semi-detached houses 20 **AA20** **External/internal walls 8**

Adjustment for external openings

Timesing	Dimension	Description
2/1/	2.40	**DDT**
	1.50	
2/2/	1.20	Half brick skin of hollow wall in facings pointed one side.
	1.05	
2/1/	2.40	
	2.10	
2/1/	1.50	
	2.10	
2/1/	1.20	&
	1.20	
2/2/	2.40	
	1.20	**DDT**
		73mm Cavity
=8 total openings		
		&
		DDT
		73mm Cavity insulation
		&
		DDT
		150mm Block skin of hollow wall in insulating blocks a.b.

Perimeter treatment

Timesing	Dimension	Description
2/1/2/	1.50	Masonry; brick/block walling; extra over walls for opening perimeters; patent plastic combined cavity closer and damp-proof barrier.
2/2/2/	1.05	
2/2/2/	2.10	
2/2+1/2/	1.20	
		NRM2 14.12.1.1

			Semi-detached houses 21	AA21		External/internal walls 9
			Lintels - ground floor			
	2/	2	Masonry; brick/block walling; proprietary and individual spot items; patent galvanised steel combined lintel and cavity tray as Catnic CG70 or similar to suit 2400mm span. NRM2 14.25.1.1.1 [LR/DR	2/	1	Galvanised steel lintel/cavity tray CG70 a.b.d. to suit 1500mm span [Hall
	2/	1	Ditto 1200mm span. [Kitchen			
2 /						

Semi-detached houses 22			AA22	External/internal walls 10
		Lintels - first floor		**NB - No deduction on blockwork for lintels**
		[BR1/2		
2/	2	Masonry; brick/block walling; proprietary and individual spot items; patent galvanised steel eaves lintel as Catnic CGE90 or similar to suit 2400mm span. NRM2 14.25.1.1.1		
		[BR3/bath		
2/	2	Ditto 1200mm span.		

			Semi-detached houses 23	AA23			External/internal walls 11
			Adjustment for internal openings				
			LB Walls				**Lintels**
							800
					2/	100	200 [Bearing
							1000
2/2+2/	0.80		**DDT**				
	2.00		100mm Dense concrete block wall a.b.				
				2/	7		Precast concrete; precast concrete goods; 1000 x 100 x 65mm pre-stressed lintel (21N/mm^2) reinforced with 1 x 12mm high tensile steel bar. NRM2 13.1.1.1
			NLB Walls				
2/1+2/	0.80		**DDT**				**[NB No deduction for blocks.**
	2.00		100mm Lightweight concrete block wall a.b.				

Semi-detached houses 24			**AA24**			**Roof Construction 1**		
		NRM2						
Section - Roof Construction								
Date 12-02-2022								**Wall plates**
Drawings	Q1/A4/2667/	1a						
		2c						
		4				Outside walls		
		5c						
		6f						12000
		7c						8000
								20000
	T.O. List						x	2
								40000
	Plates			Re-entrant	2 /		1500	3000
	Rafters							43000
	Purlins			4 /	2 /		325	2600
	Binders							
	Struts			Inside face of wall				40400
	Ceiling joists							
	Ridge			4 /	2 /		100	800
	Hip rafters							
	Valley rafters			**Outside girth of plate**				41200
						Over loadbearing partition		
						Brought forward from AA18		
								6450

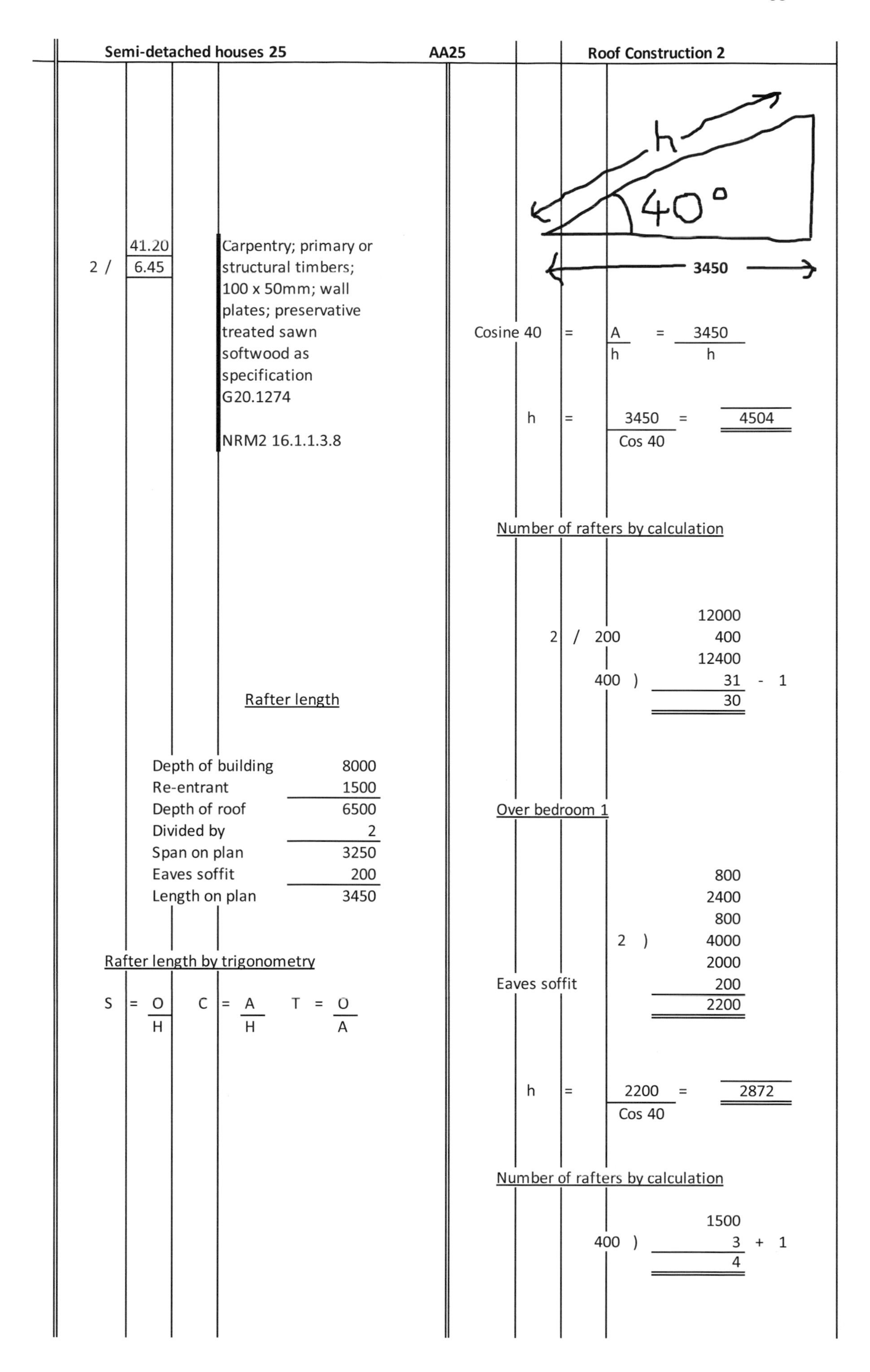
Semi-detached houses 25
AA25
Roof Construction 2
h
40°
3450
41.20
2 / 6.45
Carpentry; primary or structural timbers; 100 x 50mm; wall plates; preservative treated sawn softwood as specification G20.1274
NRM2 16.1.1.3.8
Rafter length
Depth of building 8000
Re-entrant 1500
Depth of roof 6500
Divided by 2
Span on plan 3250
Eaves soffit 200
Length on plan 3450
Rafter length by trigonometry
S = O/H C = A/H T = O/A
Cosine 40 = A/h = 3450/h
h = 3450/Cos 40 = 4504
Number of rafters by calculation
12000
2 / 200 400
12400
400) 31 - 1
30
Over bedroom 1
800
2400
800
2) 4000
2000
Eaves soffit 200
2200
h = 2200/Cos 40 = 2872
Number of rafters by calculation
1500
400) 3 + 1
4

		Semi-detached houses 26	AA26			Roof Construction 3	
30/ 2/	4.50	Carpentry; primary or structural timbers; 150 x 50mm; rafters and associated roof timbers; preservative treated sawn softwood as specification G20.1275. NRM2 16.1.1.1.8	[Main rafters			**Purlins - main roof only.**	
2/ 4/ 2/	2.87		[Over BR1				
2/	4.50		[At main hip				12000
2/	2.87		[At BR3 hip	2/	325		650 Ex wl
					Internal length		11350
				2/	11.35	Carpentry; primary or structural timbers; 200 x 100mm; purlins; preservative treated sawn softwood as specification G20.1276. NRM2 16.1.1.2.8	
						Binder - main roof	
					12.00	Carpentry; primary or structural timbers; 100 x 75mm; beams; preservative treated sawn softwood as specification G20.1277. NRM2 16.1.1.5.8	

Semi-detached houses 27 **AA27** **Roof Construction 4**

Timesing	Dimension	Description
8/ 2/	2.00	Carpentry; primary or structural timbers; 100 x 75mm; rafters and associated roof timbers; preservative treated sawn softwood as specification G20.1279. NRM2 16.1.1.1.8

Binder - over bedroom 1

Timesing	Dimension	Description
2/	1.50	Carpentry; primary or structural timbers; 100 x 50mm; beams; preservative treated sawn softwood as specification G20.1278. NRM2 16.1.1.5.8

Ceiling joists

NB Assumed jointed
Not in continuous lengths over 6m

Struts by isocoles

Rafter length	4504	
Less eaves overhang	500	(scaled
Span	4004	
Half span = strut length	2002	

Timesing	Dimension	Description	
30/	6.50	Carpentry; primary or structural timbers; 100 x 50mm; roof and floor joists; preservative treated sawn softwood as specification G20.1280. NRM2 16.1.1.4.8	[Main roof
2/ 4/	4.00		[Over BR3

Number of struts

One per 4 rafters = 4) 30

7 + 1

Struts per side of roof = 8

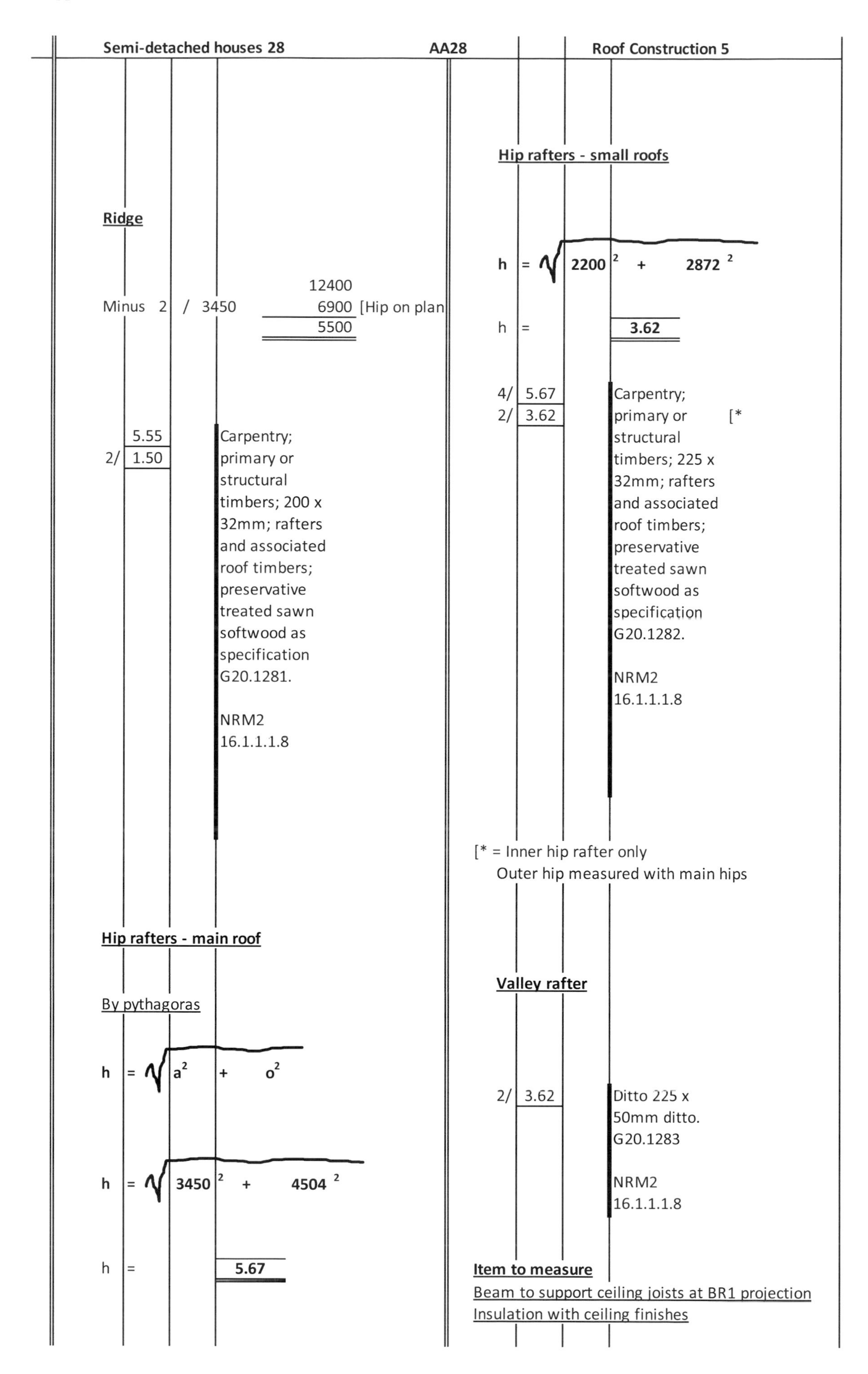

Semi-detached houses 28 AA28 **Roof Construction 5**

Ridge

			12400
Minus 2 / 3450			6900 [Hip on plan
			5500

	5.55	Carpentry; primary or structural timbers; 200 x 32mm; rafters and associated roof timbers; preservative treated sawn softwood as specification G20.1281.
2/	1.50	NRM2 16.1.1.1.8

Hip rafters - main roof

By pythagoras

$h = \sqrt{a^2 + o^2}$

$h = \sqrt{3450^2 + 4504^2}$

h = **5.67**

Hip rafters - small roofs

$h = \sqrt{2200^2 + 2872^2}$

h = **3.62**

4/	5.67	Carpentry; primary or structural timbers; 225 x 32mm; rafters and associated roof timbers; preservative treated sawn softwood as specification G20.1282. [*
2/	3.62	NRM2 16.1.1.1.8

[* = Inner hip rafter only
Outer hip measured with main hips

Valley rafter

2/	3.62	Ditto 225 x 50mm ditto. G20.1283 NRM2 16.1.1.1.8

Item to measure

Beam to support ceiling joists at BR1 projection

Insulation with ceiling finishes

Semi-detached houses 29 **AA29**

Timesing	Dimensions	Description	Note
2/	12.40	Tile and slate roof and wall coverings; plain tiling; roof coverings; 40 degrees pitch; 265 x 165mm plain sandfaced concrete tiles with 65mm minimum laps; tiles nailed every fourth course with 2 x 38 x 1.5mm aluminium alloy nails; breatheable underlay with 20 x 40mm preservative treated battens to 100mm gauge. NRM2 18.1.1.1	[Main roof
	4.50		
2/ 2/	1.50		[Over BR1
	2.87		

Roof Coverings 1

Timesing	Dimensions	Description	Note
	5.50	Tile and slate roof and wall coverings; plain tiling; boundary work; ridge; 450mm half round bedded and pointed in cement mortar (1:3); horizontal. NRM2 18.3.1.3.1	
2/	1.50		
4/	5.67	Tile and slate roof and wall coverings; plain tiling; boundary work; hips; 450mm third round bedded and pointed in cement mortar (1:3); sloping. NRM2 18.3.1.6.2	[Inner hip only to BR1 roofs]
2/	3.62		

		Semi-detached houses 30	AA30
	6	Tile and slate roof and wall coverings; plain tiling; fittings; hip iron; 250 x 6 x 38mm galvanised mild steel, screwed, base softwood. NRM2 18.4.1.4.1 & Decoration; painting and clear finishes; structural metalwork; isolated area less than or equal to 1m2; three coats of oil paint as specification. EXTERNALLY NRM2 29.3.3.2	
2/	3.62	Tile and slate roof and wall coverings; plain tiling; boundary work; valleys; purpose made tiles as specification H60123; sloping. NRM2 18.3.1.5.2	[jct small roofs with main roof]

		Roof Coverings 2
		Work to eaves
		12400 8400 20800 x 2 41600 2/ 1500 3000 44600
	44.60	Tile and slate roof and wall coverings; plain tiling; boundary work; eaves; double course of tiles nailed to battens with alloy nails; horizontal. NRM2 18.3.1.2.1 & Carpentry; timber first fixings; boarding, fascias etc; not exceeding 600mm wide; 150 x 25mm (finished); horizontal; once grooved; preservative treated wrought softwood; eaves fascia. NRM2 16.4.1.1.1+3.1 & Tile and slate roof and wall coverings; plain tiling; boundary work; eaves; horizontal; patent plastic eaves ventilator as specification. NRM2 18.3.1.2.1

Semi-detached houses 31			AA31		Roof Coverings 3
					Rainwater gutters
				45.40	Drainage above ground; rainwater installations; gutters; 100mm; straight, curved, or flexible; to timber with brass screws; PVC gutter as specification R10.10. NRM2 33.5.1.1.1.1 + p45 for background
	44.60		Carpentry; timber first fixings; boarding, fascias etc; not exceeding 600mm wide; 200 x 15mm; horizontal; exterior quality plywood; eaves soffit screwed to softwood. NRM2 16.4.1.1.1.1		
				6	Drainage above ground; gutter ancillaries; external angle. NRM2 33.6.1
			150 200 350		
				2	Drainage above ground; gutter ancillaries; internal angle. NRM2 33.6.1
	44.60 0.35		Decoration; painting and clear finishes; general surfaces >300mm girth; knot, prime, stop and paint two undercoats and one finishing coat as specification to softwood. EXTERNALLY NRM2 29.1.2.2		
				4	Drainage above ground; gutter ancillaries; running outlet. NRM2 33.6.1
				4	Drainage above ground; gutter ancillaries; balloon grating. NRM2 33.6.1
		4/	2/ 100 — 44600 800 45400		

		Semi-detached houses 32	AA32		Roof Coverings 4
		Rainwater pipes		6	Drainage above ground; rainwater installations; pipework ancillaries; 62mm connector to 100mm glazed vitrified drain; as specification R10.10.2 NRM2 33.2.1
		Height of bwk 4750 Soffit 200 Eaves 150 5100			
		Assumed 6No			
6/	5.10	Drainage above ground; rainwater installations; pipework; 62mm; straight, curved, or flexible; to masonry with brass screws; PVC rainwater pipe as specification R10.10. NRM2 33.1.1.1.1		Item	Drainage above ground; rainwater installations; marking position of and leaving or forming all holes, mortices, chases, etc. required in the structure. NRM2 33.8.1
6/	2	Drainage above ground; rainwater installations; pipework; items extra over the pipe in which they occur; fittings nominal pipe size less than 65mm; two ends; PVC rainwater pipe as specification R10.10.3 NRM2 33.3.1.2	[offset bend	Item	Drainage above ground; rainwater installations; testing and commissioning. NRM2 33.10.1

Semi-detached houses 33 **AA33**

NRM2

Section - First Floor Construction

Date 12-02-2022

Drawings Q1/A4/2667/ 1a, 2c, 6f, 7c

T.O. List

Boarding
Nosing to staircase
Joists
Trimming staircase
Strutting
Joist hangars

Boarding

Timesing	Dimension	Description	Annotation
	11.35	Carpentry; timber first fixings; boarding, flooring, sheeting etc; over 600mm wide, 19mm finished thickness; horizontal; wrought; flooring, nailed to softwood, tongued and grooved preservative treated softwood boarding in 144mm (finished) face widths as specification	
	7.35		
	DDT		
	4.65		[Re-entrant
	1.50		
2/	0.90		[Stair opg
	1.80		
	5.85		[Party wall
	0.25		
2/	6.45		[LB Ptns
	0.10		

NRM2 16.4.2.1.1.1.1

Upper floors 1

Dimension	Description
1.80	General joinery; cover fillets, stops, trims, beads, nosings etc; 40 x 19 (finished) once rounded; tongued to groove in boarding; nosing to stair opening, as specification. [Edge stair

NRM2 22.3.1.2

Joists

Dining	3350	
Ptn	100	
Kit	2100	
400)	5550	13
Add for remainder		1
Add for end joist		1
Add for ptn support		1
Total		16

		7350
Bearing	2/ 100	200
Lap at ptn		100
		7650

At re-entrant

400) 2100 = 5

At stair opening

400) 900 = 2

Trimming staircase

Timesing	Dimension	Waste	Note
		1500	[Bedroom
		100	[Partition
		1800	[Staircase
		3400	
2/	100	200	Bearing
		3600	Total length

			Semi-detached houses 34	AA34		Upper floors 2	
			Trimmer - staircase				
			Width 900				
			Bearing 100				
			Total 1000				
					2/	2	Carpentry; metal fixings, fastenings and fittings; 50 x 225 Jiffy; joist hangars; as specification [Trimmed joists]
					2/	2	NRM216.6.1.7.1
2/16/	7.65		Carpentry: primary or structural timbers; 50 x 225mm; roof and floor joists; preservative treated sawn softwood; built-in; as specification	[Main joists			
2/2/	3.60			[Trimming			
2/2/	1.00			[Trimmer			
	DDT						
2/5/	1.50			[Re-entrant			
2/2/	1.80		NRM2.16.1.1.4.8	[Stairwell			
						2	Carpentry; metal fixings, fastenings and fittings; 100 x 225 Jiffy; joist hangars; as specification [trimmer
							NRM2 16.6.1.7.1
2/2/	5.55		Carpentry: primary or structural timbers; 38 x 38mm; strutting; to softwood; preservative treated sawn softwood, double herringbone as specification	[*			
	DDT						
2/	0.90		NRM2 16.1.1.8.6	[Stair opg			
			[* Two spans per dwelling				

Semi-detached houses 35 **AA35** | **Internal finishings 1**

NRM2

Section - internal finishings
Date - 12-02-2022

Drawings Q1/A4/2667/ 1d
2c
6f
7c
9
#

T.O. List
Ceilings
Insulation
Floors
Skirtings
Walls

See finishings schedule

Ceilings

	12000	x	8000	
2/ 325	650		650	
	11350		7350	

DDT	Indent
325	
75	
1500	
850	
1500	
75	
325	
4650 x 1500	

Party wall

	8000
325	
1500	
325	2150
	5850 x 250

Partitions

LB - GF/FF

3350
900
100
2100
6450

NB - Ceiling finishings taken over NLB ptns GF/FF

Timesing	Dimensions	Description	Waste
2/	11.35	Floor, wall, ceiling,	[GF/FF
	7.35	and roof finishings;	
	DDT	finish to ceilings,	
2/	4.65	12.5mm plasterboard	[re-entrant
	1.50	and 3mm skim coat	
2/	5.85	plaster; over 600mm	[PW
	0.25	wide; to timber as	
2/ 2/	6.45	specification	[LB Ptns
	0.10		
		NRM2 28.9.2	
		&	
		Decoration; painting to general surfaces; over 300mm girth; internal; three coats of emulsion paint as specification to plastered ceilings.	
		NRM2 29.1.2.1	

N-S	8000
Re-ent	1500
Ext PW	6500

Insulation in loft

Timesing	Dimensions	Description	Waste
	12.00	Insulation, fire	
	8.00	stopping and fire	
	DDT	protection; quilts;	
	4.00	250mm thick; laid	(re-entrant
	1.50	across joists, rafters,	
	6.50	partition framing	(party wall
	0.25	or similar members	
		at 400mm centres; horizontal. Glass fibre as specification.	
		NRM2 31.3.1.2.1.	

Semi-detached houses 36 **AA36** **Internal finishings 2**

Nogging to ends of plbd

Width bldg	11350
Party wall	250
	11100

2400	)	7350
	=	3
		1
		4

Timesing	Dimension	Description
2/ 4/	11.11	Carpentry; backing and other first fix timbers; 50 x 50mm nogging; battens to softwood; sawn softwood as specification NRM2 16.3.1.2

Coving to:
Lounge
DR
Hall
Kitchen

	4000
	3350
	7350
x	2
Lounge	14700
	3250
	3350
	6600
x	2
DR	13200
	2100
	3400
	5500
x	2
Hall	11000
	2100
	2350
	4450
x	2
Kitchen	8900

Timesing	Dimension	Description	
2	14.70	Floor, wall, ceiling and roof finishings; coves; 150mm girth; standard pattern; horizontal; glued to plasterboard as specification NRM2 28.17.1.1.1+2	(LR
2	13.20		(DR
2	11.00		(Hall
2	8.90		(Kit

No deduction for paint

TT

Loft hatch with internal doors

Floors

GF

Timesing	Dimension	Description	
2/	4.00	Carpentry; boarding, flooring, sheeting, decking, casings, linings, sarking, fascias, bargeboards, soffits, etc; over 600mm wide, 20mm thick; horizontal; floors; water resistant MDF boarding floated on insulation as specification NRM2 16.4.2.1.*.1	(LR
	3.35		
2/	3.25		(DR
	3.35		
2/	2.10		(Hall
	3.40		
2/	2.10		(Kit
	2.35		

Semi-detached houses 37 **AA37**

Insulation

Timesing	Dimension	Description	Waste
2/	4.00	Insulation, fire stopping	
	3.35	and fire protection;	(LR
2/	2.10	boards; 30mm thick;	
	3.40	plain areas; horizontal;	(Hall
2/	2.10	expanded polystyrene	
	2.35	floor insulation as	(Kit
2/	3.25	specification	
	3.35		(DR
		NRM2 31.1.1.1.1	

NB Assumed partitions on GF built before floors are laid.

FF finishings with floor structure

Hardwood skirtings

Timesing	Dimension	Description	Waste
2/	14.70	General joinery;	[LR
2/	13.20	skirtings, picture	[DR
		rails; 150 x 25mm	
2/ 2/	0.20	(nominal) ovolo	(*
	DDT	pattern; mahogany	
2/	2.40	selected for clear	(DR ex dr
2/ 2/	0.80	finish to matching	(doors
		colour, plugged,	
		screwed and pellated	
		to masonry as	
		specification	
		NRM2 22.1.1.	

[*= DR ex dr reveals

&

Decoration; painting general surfaces; less than or equal to 300mm girth; internal; two coats polyurethane to hardwood as specification

NRM2 29.1.1.1

Internal finishings 3

Softwood skirtings

Bath/kitchen	8900
Hall	11000
BR1 = LR	14700
BR2 = DR	13200

BR3	1500	Land	1800
	2100		2100
	3600		3900
x	2	x	2
	7200		7800

Timesing	Dimension	Description	Waste
2/ 2/	8.90	General joinery;	(Bath/kit
2/	11.00	skirtings, picture	(Hall
2/	14.70	rails; 150 x 25	(BR1
2/	13.20	(nominal) ovolo	(BR2
2/	7.20	pattern; softwood,	(BR3
		plugged and	
2/	7.80	screwed to	(Land
2/ 2/	0.20	masonry as	(*
	DDT	specification	
2/	1.50		(Hall extl dr
2/ 4/	0.80	NRM2 22.1.1	(GF
2/ 2/ 4/	0.80		(FF

[*= Hall ex dr reveals

&

Decoration; painting general surfaces; less than or equal to 300mm girth; internal; knot, prime, stop and paint two undercoats and one finishing coat of oil paint to timber as specification

NRM2 29.1.1.1

Semi-detached houses 38 **AA38** **Internal finishings 4**

Timesing	Dimensions	Description	Location
		<u>Walls</u>	
2/ 2/	14.70 2.40	Floor, wall, ceiling and roof finishings; finish to walls, two coat render and set 12.5mm thick; over 600mm wide; to masonry; Carlite Browning render and Carlite Finish set as specification	(BR1/LR
2/ 2/	13.20 2.40		(DR/BR2
2/ 2/	8.90 2.40		(Bath/kit
2/ 1/	11.00 2.40		(Hall
2/ 1/	7.20 2.40		(BR3
2/ 1/	7.80 2.40	NRM2 28.7.2	(Land
9 rooms		&	
		Decoration, painting general surfaces; over 300mm girth; internal; three coats emulsion to plastered wall as specification NRM2 29.1.2.1	
		<u>Adjustment for wallpaper and wall tiles</u>	
2/	14.70 2.40	DDT Emulsion plastered surface as before	(LR
2/	13.20 2.40		(DR
2/	11.00 2.40		(Hall
2/	8.90 2.40		(Bath

Timesing	Dimensions	Description	Location
2/	14.70 2.40	Decoration; decorative papers or fabrics; walls and columns; areas over $1m^2$; wall paper Prime Cost Sum of £60.00/piece; to plaster as specification NRM2 29.9.2 +p11	(LR
2/	13.20 2.40		(DR
2/	11.00 2.40	Ditto £40.00/piece to plaster	(Hall
2/	8.90 2.40	Floor, wall, ceiling and roof finishings; finish to walls, ceramic tiling 150 x 150 x 4mm thick; over 600mm wide; tiling Prime Cost Sum of £30.00/m^2 supply, to plaster as specification NRM2 28.7.2 +p11	(Bath
2/	2.10 0.45	DDT Emulsion plastered surface as before	(Kit
		Assumed splashback length	
2/	2.10	Floor, wall, ceiling and roof finishings; finish to walls, ceramic tiling 150 x 150 x 4mm thick; less than or equal to 600mm wide, 450mm high; tiling Prime Cost Sum of £40.00/m^2 supply, to plaster as specification NRM2 28.7.1 +p11	

			Semi-detached houses 39	AA39			Internal finishings 5
							(Metal lathing to plates - ff
			Adjustments				
					2/	11.10	Floor, wall, ceiling and roof
					2/	7.35	finishings; metal mesh
			Wall finishes extl opgs		2/ 2/	6.45	lathing; to walls; 50mm (*
							wide, nailed with
							galvanised nails; to timber;
2/ 1/	2.40		DDT	(LR			galvanised steel as
	1.50		Render and set to				specification
2/ 2/	1.20		masonry walls as	(Kit/bath			
	1.05		before				NRM2 28.31.1.1.1+2
2/ 1/	2.40			(DR			
	2.10						(* Both sides of LB partition
2/ 1/	1.50			(Hall			
	2.10						Beads
2/ 1/	1.20			(BR3			
	1.20				2/ 4/	2.40	Floor, wall, ceiling (LR/DR/BR1/2
2/ 2/	2.40			(BR1/2	2/ 3/	1.20	and roof (Kit/bath/BR3
	1.20				2/ 1/	1.50	finishings; beads, (Hall
					2/1/ 2/	1.50	angle; standard (LR
					2/2/ 2/	1.05	plaster bead; on (Kit/bath
				(Reveals	2/2/ 2/	2.10	plaster dabs to (DR/hall
					2/3/ 2/	1.20	masonry; stainless (BR 1/2/3
2/ 1/ 2/	1.50		Floor, wall, ceiling	(LR			steel as
2/ 2/ 2/	1.05		and roof finishings;	(Kit/bath			specification
2/ 2/ 2/	2.10		finish to walls; two	(DR/hall			
2/ 3/ 2/	1.20		coat render and set	(BR 1/2/3			NRM2 28.28.1
			12.5mm thick; less				
			than or equal to		2/ 1/	1.20	**DDT** (Kit
			600mm wide,			1.05	Three coat
			200mm wide; to		2/ 1/	1.20	emulsion to (BR3
			masonry as before			1.20	plaster as before
			described as		2/ 2/	2.40	(BR1/2
			specification			1.20	
							(Reveals
2/ 4/	2.40		Floor, wall, ceiling	(a			
2/ 3/	1.20		and roof finishings;	(b			
2/ 1/	1.50		finish to ceilings, two	(c	2/ 2/	1.20	**LESS** (Soffits
			coat render and set			0.20	Last
			12.5mm thick; less		2/ 2/	2.40	(ditto
			than or equal to			0.20	
			600mm wide,		2/ 2/	1.05	(Jambs
			200mm wide; to			0.20	
			keyed steel lintel,		2/ 6/	1.20	(ditto
			Carlite Browning			0.20	
			render and Carlite				
			Finish set as				
			specification				
			NRM2/28.9.1				
			(a) (LR/DR/BR1/2)				
			(b) (Kit/bath/BR3				
			(c) (Hall				
Tot = 8 openings							

		Semi-detached houses 40	AA40			Internal finishings 6	
2/	2.40 1.50	**DDT** Decorative paper PC £60.00/piece to walls, area >1m2 as before	(LR	2/	1.20 1.05	DDT Ceramic tiling; walls > 600mm wide, £30.00/m2 as before described	(Bath
2/	2.40 2.10		(DR				
			(Reveals				
2/ 2/	2.40 0.20	**LESS** Last	(head				
2/ 2/	1.50 0.20		(jambs				
2/ 2/	2.10 0.20		(Ditto				
							(Reveals
				2/	1.20	Floor, wall, ceiling and roof finishings; finish to ceilings, ceramic tiling 150 x 150 x 4mm thick; less than or equal to 600mm wide, 200mm wide; tiling Prime Cost Sum of £30.00/m2 supply, to plaster as specification NRM2 28.9.1 +p11	(Bath
2/	1.50 2.10	DDT Decorative paper PC £40.00/piece to walls, area >1m2 as before	(Hall				
			(Reveals				
2/	1.50 0.20	LESS Last	(head	2/ 2/	1.05	Floor, wall, ceiling and roof finishings; finish to walls, ceramic tiling 150 x 150 x 4mm thick; less than or equal to 600mm wide, 200mm wide; tiling Prime Cost Sum of £30.00/m2 supply, to plaster as specification NRM2 28.7.1	(Bath
2/ 2/	2.10 0.20		(jambs				

Semi-detached houses 41 **AA41**

Timesing	Dimension	Squaring	Description
2/ 1/ 2/ 2/	1.20 1.05		Floor, wall, ceiling and roof finishings; special tiles, slabs or blocks; rounded edge tile, Tiling PC £30.00/m2 as before, as specification (Bath wdw) NRM2 28.23.1 +p11
2/	2.10 0.45		Floor, wall, ceiling and roof finishings; special tiles, slabs or blocks; rounded edge tile, Tiling PC £40.00/m2 as before, as specification (Kit NRM2 28.23.1 +p11
	2		Floor, wall, ceiling and roof finishings; special tiles, slabs or blocks; REX tile, Tiling PC £40.00/m2 as before, as specification (Kit NRM2/28.23.1

Internal finishings 7

Internal door adjustments

Timesing	Dimension	Squaring	Description
2/ 7/ 2/	0.80 2.00		**DDT** (All intl doors Render and set masonry walls as before & **DDT** Three coats emulsion paint as before
2/ 2/	0.80 2.00		**DDT** (DR/LR Decorative paper PC £60.00/piece to walls as before & **ADD** Three coats emulsion paint as before

			Semi-detached houses 42	AA42 Internal finishings 8
2/ 3/	0.80 2.00		**DDT** (hall Decorative paper PC £40.00/piece to walls as before **ADD** Three coats emulsion paint as before	**NB No adjustments made for decorative finishings behind skirtings.**
2	0.80 2.00		**DDT** (bath Ceramic tiling; walls to plaster > 600 wide, PC £30.00/m2 as before & **ADD** Three coats emulsion paint as before	

			Semi-detached houses 43	AA43			Windows and doors 1
			External windows/doors				
	Item		Allow the Defined Provisional Sum of £16000.00 for UPVC factory glazed windows and doors to be supplied by an approved manufacturer, including building in templates and plugging and screwing with stainless steel screws to masonry base. General configuration as drawings Q1/A4/2667/8-11 Defined Provisional Sum as NRM2 Glossary, Definitions, p7	2/	1		**FIX** (Hall Windows and window frames; 1500 x 2100 Door/sidelight unit NRM2 23.1.1.*.1+2 NRM2 24.1.1+2.1
				2/	1		**FIX** (BR3 Windows and window frames; 1200 x 1200 Window NRM2 23.1.1.*.1+2
2/	1		**FIX** (LR Windows and window frames; 2400 x 1500 window as specification. NRM2 23.1.1.*.1+2	2/	2		**FIX** (BR1,2 Windows and window frames; 2400 x 1200 Window NRM2 23.1.1.*.1+2
							End of fix items
				Window boards			
				Stooling		2/75	2400 + 150 = 2550; 1200 + 150 = 1350
2/	2		**FIX** (kit/bath Windows and window frames; 1200 x 1050 window NRM2 23.1.1.*.1+2	2/ 3/ 2/ 3/	2.55 1.35		General joinery; (LR,BR1+2 window boards; (K,Bath,BR3 225 x 32, rounded front edge and tongue to rear edge, plugged and screwed to masonry; MDF as specification NRM2 22.5.1+2
2/	1		**FIX** (DR Windows and window frames; 2400 x 2100 French door NRM2 24.1.1+2.1				

			Semi-detached houses 44		AA44			Windows and doors 2
			(Wdw bds					
2/ 3/	2.55		Decoration; painting to general surfaces; less than or equal to 300mm girth; internal; knot, prime, stop and paint two undercoats and one finishing coat of oil paint to timber as specification NRM2 29.1.1.1	2/ 7/	0.80			Door, shutters and hatches; door stops; 50 x 15 (finished); plain; planted on, as specification NRM2 24.11.1.1
2/ 3/	1.35			2/ 7/ 2/	2.00			
								2000
								x 2
			Internal doors					4000
						Arch		## 800
						lap	-	## 4800
						tot	2/ 2/	## 200
								5000
2/	7		Door, shutters and hatches; doors; 38 x 736 x 1968 plywood faced flush door as specification NRM2 24.2.1					
				2/ 7/ 2/	5.00			General joinery; architraves etc.; 60 x 16mm (Finished) softwood architrave, splayed and rounded as specification (Quad NRM2 22.2.1+2
					DDT			
				2/12/	2.00			
								5
								7
2/ 7/	0.80		Door, shutters and hatches; door linings; 125 x 32 (finished); plain; plugged and screwed to masonry, as specification NRM2 24.10.1.1					(* where quadrant 12
2/ 7/ 2/	2.00							
				2/12/	2.00			General joinery; cover fillets, stops, trims, beads, nosings etc.; 20 x 20mm (Finished) softwood, quadrant as specification (Quad NRM2 22.3.1
								Door paint area
							Face	1968 736
							Edge	38 38
								2006 x 774
								Say 2000 x 800

			Semi-detached houses 45	AA45			Windows and doors 3
			Doors - Paint		2/	7	Doors, shutters and hatches; ironmongery; internal door latch screwed to timber as specification NRM2 24.16.1+2.1-3
2/ 7/ 2/	0.80 2.00		Decoration; painting to general surfaces; over 300mm girth; internal; prime, stop and paint two undercoats and one finishing coat of oil paint to plywood as specification NRM2 29.1.2.1				
			Girth **Linings** lining 125 edge lining 2/ ## 64 arch 2/ ## 120 arch edge 4/ ## 64 stop edges 2/ ## 30 **less** arch o/lap 2/ ## -20 Total girth 383 (NB no adjustment where quadrant used instead of architrave		2/	7	Doors, shutters and hatches; ironmongery; internal door furniture screwed to timber as specification NRM2 24.16.1+2.1-3
							Hatch
					2/	1	Door, shutters and hatches; hatches; 38 x 750 x 750 plywood faced flush hatch as specification NRM2 24.6.1
2/ 7/ 1/ 2/ 7/ 2/	0.80 0.38 2.00 0.38		Decoration; painting to general surfaces; over 300mm girth; internal; knot, prime, stop and paint two undercoats and one finishing coat of oil paint to timber as specification NRM2 29.1.2.1			Lining	4/ 750 3000 (lining
			Ironmongery				
2/ 7/	1		Doors, shutters and hatches; ironmongery; PAIR 100 x 100 pressed steel butts screwed to timber as specification NRM2 24.16.1+2.1-3		2/	3.00	Door, shutters and hatches; door linings; 75 x 25 (finished); plain; screwed to timber, as specification NRM2 24.10.1.1

Semi-detached houses 46 **AA46** **Windows and doors 4**

Timesing	Dimension	Squaring	Description
			Lining 3000 Corners 4/ 2/ 60/ 480 3480
2/	3.48		General joinery; architraves etc.; 60 x 16mm (Finished) softwood architrave, splayed and rounded as specification NRM2 22.2.1
			750 Edge 38 788 say 800
2/ 2/	1		Decoration; painting to general surfaces; isolated area 800 x 800mm; internal; prime, stop and paint two undercoats and one finishing coat of oil paint to plywood as specification NRM2 29.1.3.1
	3.48		Decoration; painting to general surfaces; less than or equal to 300mm girth; internal; knot, prime, stop and paint two undercoats and one finishing coat of oil paint to timber as specification NRM2 29.1.1.1

Timesing	Dimension	Squaring	Description
			(trimmers
2/ 2/	0.75		Carpentry; primary or structural timbers; 100 x 50mm; roof and floor joists; preservative treated sawn timber as before as specification NRM2 16.1.1.4
			(No ddn for trimmed joist
2/ 2/	6		Carpentry; metal fixings, fastenings and fittings; 50 x 100 Jiffy; joist hangar; nailed to timber; as specification NRM2 16.6.1.7.1

			Semi-detached houses 47	AA47			Internal fittings 1	
			Staircase		2/	0.90 1.80	**DDT**	(Stairwell
					2/	0.90 1.80	Decoration; painting to general surfaces; over 300mm girth; internal; three coats of emulsion paint to ceiling as before.	(Over S/C
	Item		Allow the Defined Provisional Sum of £7500.00 for Supply and fitting of purpose made timber staircase consisting of thirteen steps and twelve risers, newels, handrails and balustrading including three kite winders, to general configuration as shown on drawings Q1/A4/2667 1a and Q1/A4/2667 2c to be supplied and fitted by an approved specialist. Defined Provisional Sum as NRM2 Glossary, Definitions, p7				**(First floor adjs in floor const**	
					2/	0.90 1.20	Floor, wall, ceiling, and roof finishings; finish to ceilings, 12.5mm plasterboard and 3mm skim coat plaster; over 600mm wide; over 3.50m above structural floor level to timber as specification NRM2 28.9.2.1	(Over S/C (say
			Builders work in connection				&	
2/	0.90 1.80		**DDT**	(Stairwell			Decoration; painting to general surfaces; over 300mm girth; internal; over 3.5m but not exceeding 5m above finished floor level; three coats of emulsion paint as specification to plastered ceilings. NRM2 29.1.2.1.1	
2/	0.90 1.20		Floor, wall, ceiling, and roof finishings; finish to ceilings, 12.5mm plasterboard and 3mm skim coat plaster; over 600mm wide as before	(Over S/C (say				

			Semi-detached houses 48		AA48
			1800	(S/C side	
			900	(S/C end	
			2700		
	2.70		**DDT** Floor, wall, ceiling and roof finishings; coves; 150mm girth; standard pattern as before	(Cove (to S/C	
			Underside of staircase		
2/	0.90 2.40		Floor, wall, ceiling, and roof finishings; finish to ceilings, 12.5mm plasterboard and 3mm skim coat plaster; over 600mm wide as before; to sloping soffit of staircase. **PROVISIONAL QUANTITY** NRM2 28.9.2 + 3.2.7, page 47		
2/ 3/ 2/	1.20		Ditto; less than or equal to 600mm wide average 300mm wide; to sides or soffit of staircase tread. **PROVISIONAL QUANTITY** NRM2 28.9.1	(Under (stairs	

			Internal fittings 2	
2/	0.90 2.40		Three coats of emulsion paint to ceilings; as before. **PROVISIONAL QUANTITY** NRM2 3.2.7, page 47	(U/S S/C
2/ 3/ 2/	1.20 0.30			(U/S (Trds (& risers
			Edge of opening	
			225	(Joist
			25	(projn
			250	
2/	1.80		General joinery; Cover fillets, stops, trims, beads, nosings, etc.; 250 x 32mm (finished) edge of staircase opening, with rounded bottom edge; MDF as specification. NRM2 22.3.1+2	(staircase (trim
			250 32 25 **307**	
2/ 2/	1.80 1.80		Decoration; painting to general surfaces; over 300mm girth; internal; knot, prime, stop and paint two undercoats and one finishing coat of oil paint to timber as before. **PROVISIONAL QUANTITY** NRM2 29.1.2.1 + 3.2.7, page 47	(staircase
2/	0.90 2.40			(treads/ risers
2/	1.80 0.31			(Edge S/C

			Semi-detached houses 49	AA49			Internal fittings 3
			Balustrade - assumed 100) 2400 **24**				
2/ ### / 2 /	0.80 2.40		Decoration; painting to general surfaces; not exceeding 300mm girth; internal; knot, prime, stop and paint two undercoats and one finishing coat of oil paint to timber as before. **PROVISIONAL QUANTITY** NRM2 29.1.2.1 + 3.2.7, page 47	(Balust (Handr			**TT** ~~Kitchen fittings~~ ~~Bathroom fittings~~
					Item		Furniture, fittings and equipment; Allow the Defined Provisional Sum of £5000.00 for Supply and fitting of standard kitchen fittings to be supplied and fitted by an approved specialist. Defined Provisional Sum as NRM2 Glossary, Definitions, p7
	Item		Allow the Undefined Provisional Sum of £500.00 for additional builder's work in connection with the timber staircase as shown on drawings Q1/A4/2667 1a and Q1/A4/2667 2c. Undefined Provisional Sum as NRM2 Glossary, Definitions, p12				
					Item		Furniture, fittings and equipment; Allow the Defined Provisional Sum of £3000.00 for Supply and fitting of standard bathroom fittings to be supplied and fitted by an approved specialist. Defined Provisional Sum as NRM2 Glossary, Definitions, p7

			Semi-detached houses 50	AA50			Internal fittings 4
	Item		Furniture, fittings and equipment; Allow the Undefined Provisional Sum of £800.00 for builder's work in connection with kitchen and bathroom fittings Undefined Provisional Sum as NRM2 Glossary, Definitions, p12				

Appendix 2

Bills of quantities – measured works

		Bills of Quantities For Semi-detached houses				
		Measured work				
						£.p
		Substructure				
		Excavating and Filling				
		Site preparation				
	1	Remove topsoil 150mm deep	97	m^2		
		Excavation				
	2	Excavation starting 150mm below existing ground level; foundation excavation not exceeding 2.0m deep	29	m^3		
		Support to faces of excavation				
	3	Maximum depth less than or equal to 1.00m; trench foundations; distance between opposing faces less than or equal to 2.00m.	105	m^2		
		Disposal				
	4	Excavated material off-site.	27	m^3		
		Filling obtained from excavated material				
	5	150mm thick; topsoil, 20m distant, rotovated and selected.	97	m^2		
	6	450mm thick; backfilling foundations in 200mm layers as specification.	7	m^2		
		Imported filling				
	7	Beds and voids over 50mm thick but not exceeding 500mm thick, 450mm thick; level; backfilling to foundations; hardcore as specification compacted in layers.	5	m^3		

	8	Beds and voids over 50mm thick but not exceeding 500mm thick, 150mm thick; level; over-site bed; hardcore as specification.	11	m^3		
		Damp proof membrane				
	9	Over 500mm wide, 2.7 micron; horizontal polythene lapped 300mm at joints on blinded hardcore.	85	m^2		
		Insitu concrete works				
	10	Plain insitu concrete; horizontal work; over 300mm thick; in structures; poured on or against earth or unblinded hardcore; 15/N/mm2 - 40mm aggregate.	14	m^3		
	11	Plain insitu concrete; horizontal work less than or equal to 300mm thick; in structures; 21N/mm2 - 20mm aggregate.	11	m^3		
	12	Vertical work; less than or equal to 300mm thick; in structures; concrete 21N/mm2 - 10mm aggregate.	2	m^3		
	13	Plain insitu concrete; tamping by mechanical means; top surface.	74	m^2		
		Masonry				
		Brick/block walling				
		Walls; blockwork; stretcher bond; dense concrete (10.5 N/mm2) blocks in cement mortar (1:4).				
	14	100mm wall.	10	m^2		
	15	100mm Skins of hollow wall	28	m2		
	16	150mm skins of hollow wall.	31	m^2		

		Walls; Brickwork; Butterly Brown Brindle bricks manufactured by the Butterly Brick Company, bedded, jointed and pointed in cement mortar (1:4) in stretcher bond, pointed with a neat weatherstruck joint as the work proceeds.				
	17	Skins of hollow walls; half brick thick; facework one side	13	m^2		
		Forming cavities; stainless steel wall ties spaced 600 horizontally and vertically				
	18	50mm wide	5	m^2		
	19	75mm wide	31	m^2		
		Damp proof courses less than or equal to 300mm wide; single layer vinyl as specification.				
	20	Horizontal.	108	m		
		TOTAL SUBSTRUCTURE				

		Superstructure				
		Precast concrete				
		Precast concrete goods				
	21	1000 x 100 x 65mm pre-stressed lintel (21N/mm2) reinforced with 1 x 12mm high tensile steel bar.	14	No		
		Masonry				
		Brick/block walling				
		Walls; blockwork; stretcher bond; dense concrete blocks (10.5N/mm2) in gauged mortar (1:1:6).				
	22	100mm wall	56	m^2		
	23	100mm skins of hollow wall.	77	m^2		
		Walls; blockwork; stretcher bond; lightweight concrete blocks in gauged mortar (1:1:6).				
	24	100mm wall	47	m^2		
		Walls; blockwork; stretcher bond; insulating concrete blocks in gauged mortar (1:1:6).				
	25	150mm skins of hollow wall.	162	m^2		
		Walls; Brickwork; Butterly Brown Brindle bricks manufactured by the Butterly Brick Company, bedded, jointed and pointed in gauged mortar (1:1:6) in stretcher bond, pointed with a neat weatherstruck joint as the work proceeds.				
	26	Skins of hollow walls; half brick thick; facework one side	159	m^2		
		Extra over walls for opening perimeters;				
	27	Patent plastic combined cavity closer and damp-proof barrier.	46	m		
		Forming cavities; stainless steel wall ties spaced 600 horizontally and vertically				
	28	50mm wide	40	m^2		
	29	73mm wide	156	m^2		

		Cavity insulation; glass-fibre resin treated wall bats, fixed between wall ties.				
	30	73mm Thick	167	m^2		
		Masonry; brick/block walling; proprietary and individual spot items;				
		Patent galvanised steel combined lintel and cavity tray as Catnic CG70 or similar				
	31	1200mm Span	2	No		
	32	1500mm Span	2	No		
	33	2400mm span.	4	No		
		Patent galvanised steel eaves lintel as Catnic CN55A or similar.				
	34	1200mm Span	4	No		
	35	2400mm span.	4	No		
		Carpentry				
		Primary or structural timbers preservative treated sawn softwood as specification G20.1274-1288.				
	36	200 x 32mm; rafters and associated roof timbers; .	9	m		
	37	225 x 32mm; rafters and associated roof timbers; .	30	m		
	38	150 x 50mm; rafters and associated roof timbers; .	331	m		
	39	225 x 50mm; rafters and associated roof timbers; .	7	m		
	40	100 x 75mm; rafters and associated roof timbers; .	32	m		
	41	200 x 100mm Purlin	23	m		
	42	100 x 50mm Wall plate	54	m		
	43	100 x 50mm Roof and floor joists	230	m		
	44	225 x 50mm Roof and floor joists	241	m		
	45	100 x 50mm Beams	3	m		

Taking Off Domestic Building Construction - Bills of Quantities page BQ5

	46	100 x 75 Beams	12	m		
	47	38 x 38mm; Double herringbone strutting.	20	m		
		Carpentry; backing and other first fix timbers;				
	48	50 x 50mm nogging; battens to softwood; sawn softwood as specification	89	m		
		Preservative treated ; boarding, flooring, sheeting, decking, casings, linings, sarking, fascias, bargeboards, soffits, etc;				
		Not exceeding 600mm wide;				
	49	150 x 25mm (finished); Wrought softwood horizontal; once grooved; eaves fascia.	45	m		
	50	200 x 15mm; Exterior quality plywood; horizontal; eaves soffit screwed to softwood	45	m		
		Over 600mm wide,				
	51	19mm (finished); Wrought softwood; horizontal; flooring, tongued and grooved softwood boarding in 144mm (finished) face widths as specification	70	m^2		
	52	20mm; Water resistant MDF; horizontal; floor boarding floated on insulation as specification	73	m^2		
		Metal fixings, fastenings and fittings as specification;				
	53	50 x 100 Jiffy; joist hangers;	12	No		
	54	50 x 225 Jiffy; joist hangers;	8	No		
	55	100 x 225 Jiffy; joist hangers;	2	No		

		Tile and slate roof and wall coverings;				
		Plain tiling; 40 degrees pitch; 265 x 165mm plain sandfaced concrete tiles with 65mm minimum laps; tiles nailed every fourth course with 2 x 38 x 1.5mm aluminium alloy nails; breatheable underlay with 20 x 40mm preservative treated battens to 100mm gauge.				
	56	Roof coverings;	129	m^2		
		Boundary work;				
	57	Eaves; horizontal; double course of tiles nailed to battens with alloy nails.	45	m		
	58	Eaves; horizontal; patent plastic eaves ventilator as specification.	45	m		
	59	Ridge; horizontal; 450mm half round bedded and pointed in cement mortar (1:3).	9	m		
	60	Hips; sloping; 450mm third round bedded and pointed in cement mortar (1:3).	30	m		
	61	Valleys; sloping; purpose made tiles as specification.	7	m		
		Fittings;				
	62	Hip iron; 250 x 6 x 38mm galvanised mild steel, screwed, base softwood.	6	No		
		General joinery				
		Skirtings, picture rails;				
	63	150 x 25 (nominal) Ovolo pattern; softwood, plugged and screwed to masonry as specification	122	m		
	64	150 x 25mm (nominal) Ovolo pattern; mahogany selected for clear finish to matching colour, plugged, screwed and pellated to masonry as specification	49	m		
		Architraves etc.; as specification.				
	65	60 x 16mm (Finished) softwood architrave, splayed and rounded	99	m		

		Cover fillets, stops, trims, beads, nosings etc; as specification.				
	66	20 x 20mm (Finished) softwood, quadrant.	48	m		
	67	40 x 19 (finished) once rounded; tongued to groove in boarding; softwood nosing to stair opening.	2	m		
	68	250 x 32mm (finished) once rounded; edge of staircase opening, MDF.	4	m		
		Window boards; plugged and screwed to masonry; rounded front edge and tongue to rear edge; MDF as specification.				
	69	225 x 32 Board	23	m		
		Windows, screens and lights				
	70	Allow the Defined Provisional Sum of £16000.00 for UPVC factory glazed windows and external doors to be supplied by an approved manufacturer, including building in templates and plugging and screwing with stainless steel screws to masonry base. General configuration as drawings Q1/A4/2667/8-11	Item			
		FIX the following items;				
		Windows and window frames;				
	71	1200 x 1050 Window	4	No		
	72	1200 x 1200 Window	2	No		
	73	2400 x 1200 Window	4	No		
	74	2400 x 1500 Window	2	No		
	75	1500 x 2100 Door/sidelight unit	2	No		
	76	2400 x 2100 French door	2	No		
		Door, shutters and hatches;				
		Plywood faced flush door as specification				
	77	38 x 736 x 1968 Door	14	No		

	78	38 x 750 x 750 Hatch.	2	No		
		Door linings; as specification				
	79	75 x 25 (finished); plain; screwed to timber.	6	m		
	80	125 x 32 (finished); plain; plugged and screwed to masonry.	67	m		
		Door stops; planted on, as specification				
	81	50 x 15 (finished); plain;	67	m		
		Ironmongery; screwed to timber as specification.				
	82	PAIR 100 x 100 pressed steel butts;	14	No		
	83	Internal door latch;	14	No		
	84	Internal door furniture	14	No		
		Stairs, walkways and balustrades				
	85	Allow the Defined Provisional Sum of £7500.00 for Supply and fitting of purpose made timber staircase consisting of thirteen steps and twelve risers, newels, handrails and balustrading including three kite winders, to general configuration as shown on drawings Q1/A4/2667 1a and Q1/A4/2667 2c to be supplied and fitted by an approved specialist.	Item			
	86	Allow the Undefined Provisional Sum of £500.00 for additional builder's work in connection with the timber staircase as shown on drawings Q1/A4/2667 1a and Q1/A4/2667 2c.	Item			
		Floor, wall, ceiling and roof finishings;				
		Two coat render and set 12.5mm thick; to masonry; Carlite Browning render and Carlite Finish set as specification				
		Finish to walls,				
	87	Over 600mm wide.	390	m^2		
	88	Less than or equal to 600mm wide, 200mm wide	46	m		

		Ceramic tiling 150 x 150 x 4mm thick; tiling Prime Cost Sum of £30.00/m2 supply, to plaster as specification				
		Finish to walls,				
	89	Over 600mm wide;	37	m^2		
	90	Less than or equal to 600mm wide, 200mm wide;	4	m		
	91	Rounded edge tile	7	m		
		Finish to ceilings,				
	92	Less than or equal to 600mm wide, 200mm wide	2	m		
		Ceramic tiling 150 x 150 x 4mm thick; tiling Prime Cost Sum of £40.00/m2 supply, to plaster as specification				
	93	Less than or equal to 600mm wide, 450mm wide.	4	m		
	94	Roundod edge tile	6	m		
	95	REX Rounded corner tile	4	No		
		12.5mm plasterboard and 3mm skim coat plaster; to timber as specification				
		Finish to ceilings,				
	96	Over 600mm wide;	142	m^2		
	97	Over 600mm wide; over 3.50m above structural floor level	2	m2		
	98	Over 600mm wide; to sloping soffit of staircase. **PROVISIONAL**	4	m2		
	99	Less than or equal to 600mm wide average 300mm wide; to sides or soffit of staircase tread. **PROVISIONAL**	14	m		
		Two coat render and set 12.5mm thick; to keyed steel lintel, Carllte Browning render and Carlite Finish set as specification				
		Finish to ceilings,				
	100	Less than or equal to 600mm wide, 200mm wide;	29	m		

		Coves; standard pattern; horizontal; glued to plasterboard as specification				
	101	150mm girth;	93	m		
		Beads, angle; on plaster dabs to masonry; stainless steel as specification				
	102	Standard plaster bead;	75	m		
		Metal mesh lathing; galvanised steel as specification				
	103	To walls; 50mm wide, nailed with galvanised nails; to timber;	63	m		
		Decoration.				
		EXTERNALLY				
		Painting and clear finishes;				
		Knot, prime, stop and paint two undercoats and one finishing coat as specification to timber.				
	104	General surfaces greater than 300mm girth.	16	m^2		
		Painting structural metalwork				
		Three coats of oil paint as specification.				
	105	Isolated area less than or equal to 1m2.	6	No		
		INTERNALLY				
		Painting and clear finishes;				
		Three coats of emulsion paint as specification				
	106	General surfaces; over 300mm girth; to plastered ceilings;	142	m^2		
	107	Ditto **PROVISIONAL**	9	m2		
	108	General surfaces; over 300mm girth; over 3.5m but not exceeding 5m above finished floor level; to plastered ceilings;	2	m^2		
	109	General surfaces; over 300mm girth; to plastered walls;	211	m^2		

		Prime, stop and paint two undercoats and one finishing coat of oil paint to plywood as specification				
	110	General surfaces; over 300mm girth;	45	m2		
	111	General surfaces; isolated area 800 x 800mm;	4	No		
		Knot, prime, stop and paint two undercoats and one finishing coat as specification to timber.				
	112	General surfaces over 300mm girth	26	m2		
	113	General surfaces over 300mm girth **PROVISIONAL**	18	m2		
	114	General surfaces less than or equal to 300mm girth.	149	m		
	115	Ditto; **PROVISIONAL**	43	m		
		Two coats polyurethane to hardwood as specification				
	116	General surfaces less than or equal to 300mm girth.	49	m		
		Decorative papers or fabrics; to plaster as specification				
	117	Walls and columns; areas over 1m2; paper Prime Cost Sum of £40.00/piece;	39	m2		
	118	Walls and columns; areas over 1m2; paper Prime Cost Sum of £60.00/piece;	115	m2		
		Insulation, fire stopping and fire protection				
		Expanded polystyrene floor insulation as specification				
	119	Boards; 30mm thick; plain areas; horizontal;	73	m2		

		Glass fibre as specification;				
	120	Quilts; 250mm thick; laid across joists, rafters, partition framing or similar members at 400mm centres; horizontal.	88	m2		
		Furniture, fittings and equipment;				
	121	Allow the Defined Provisional Sum of £5000.00 for Supply and fitting of standard kitchen fittings to be supplied and fitted by an approved specialist.	Item			
	122	Allow the Defined Provisional Sum of £3000.00 for Supply and fitting of standard bathroom fittings to be supplied and fitted by an approved specialist.	Item			
	123	Allow the Undefined Provisional Sum of £800.00 for builder's work in connection with kitchen and bathroom fittings	Item			
		Drainage above ground;				
		Rainwater installations;				
		PVC rainwater pipes to masonry with brass screws as specification R10.10.				
	124	62mm; straight, curved, or flexible.	31	m		
	125	Extra for fittings; two ends as specification R10.10.3	12	No		
		Pipework ancillaries;				
	126	62mm connector to 100mm glazed vitrified drain; as specification R10.10.2	6	No		
		PVC gutters; to timber with brass screws; as specification R10.10.				
	127	100mm; straight, curved, or flexible.	45	m		
		Gutter ancillaries;.				
	128	External angle	6	No		
	129	Internal angle	2	No		
	130	Running outlet	4	No		
	131	Baloon grating	4	No		

Taking Off Domestic Building Construction - Bills of Quantities page BQ13

		Marking position of and leaving or forming all holes, mortices, chases, etc. required in the structure.				
	132	Rainwater installations;	Item			
		Testing and commissioning				
	133	Rainwater installations;	Item			
		TOTAL SUPERSTRUCTURE				

Appendix 3

Drawings

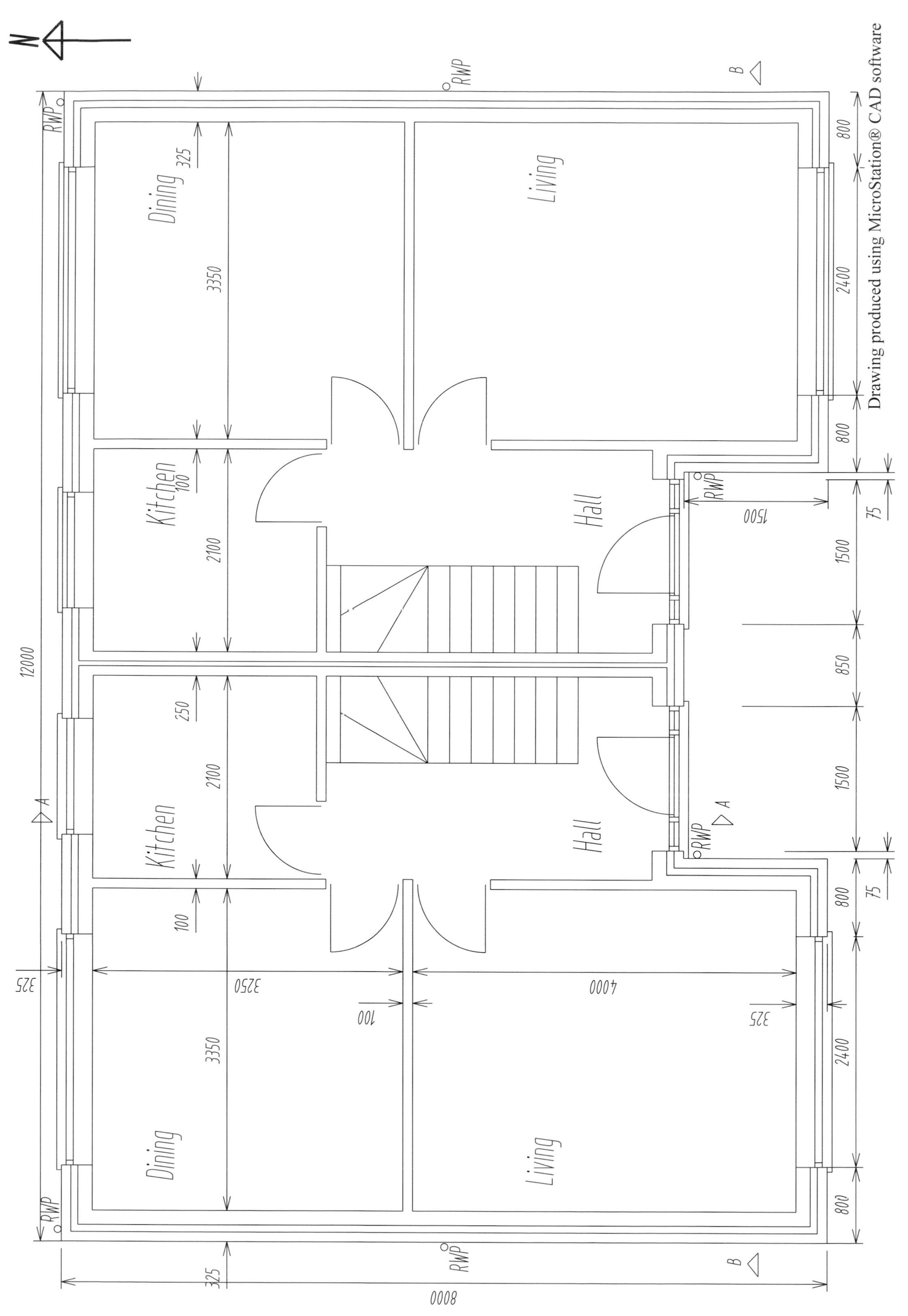

Andrew R Atkinson PhD MSc FRICS CertEd. Semi-Detached Houses. Drawing Q1/A4/2667-1a. Date 01-08-2023, Scale 1:50
GROUND FLOOR PLAN

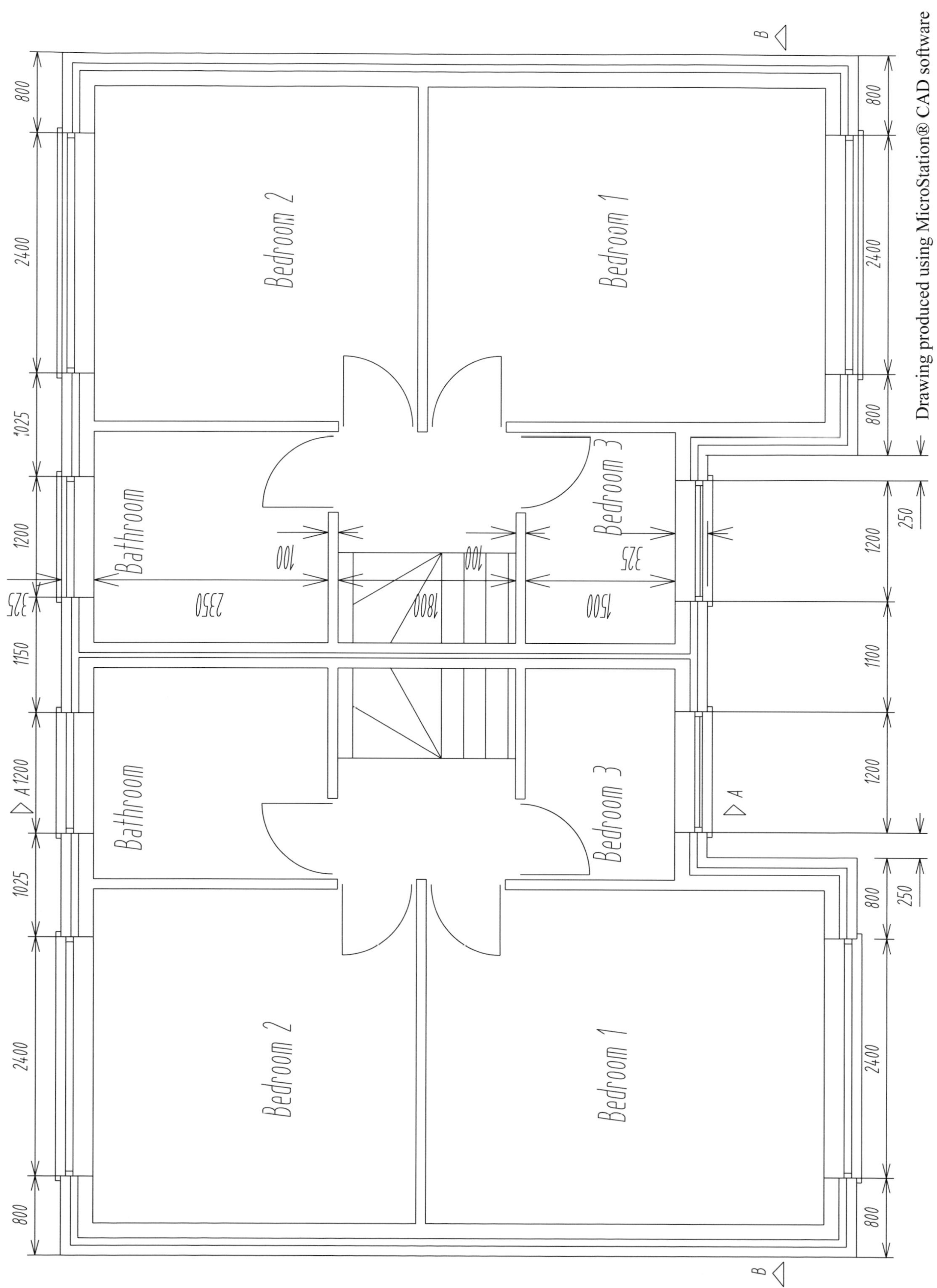

Andrew R Atkinson PhD MSc FRICS CertEd. Semi-Detached Houses. Drawing Q1/A4/2667-2c. Date 01-08-2023, Scale 1:50
FIRST FLOOR PLAN

Andrew R Atkinson PhD MSc FRICS CertEd. Semi-Detached Houses. Drawing Q1/A4/2667-3a. Date 01-08-2023, Scale 1:50

FOUNDATION PLAN

Andrew R Atkinson. Semi-Detached Houses. Drawing Q1/A4/2667-4. Date 01-08-2023, Scale 1:50
ROOF PLAN

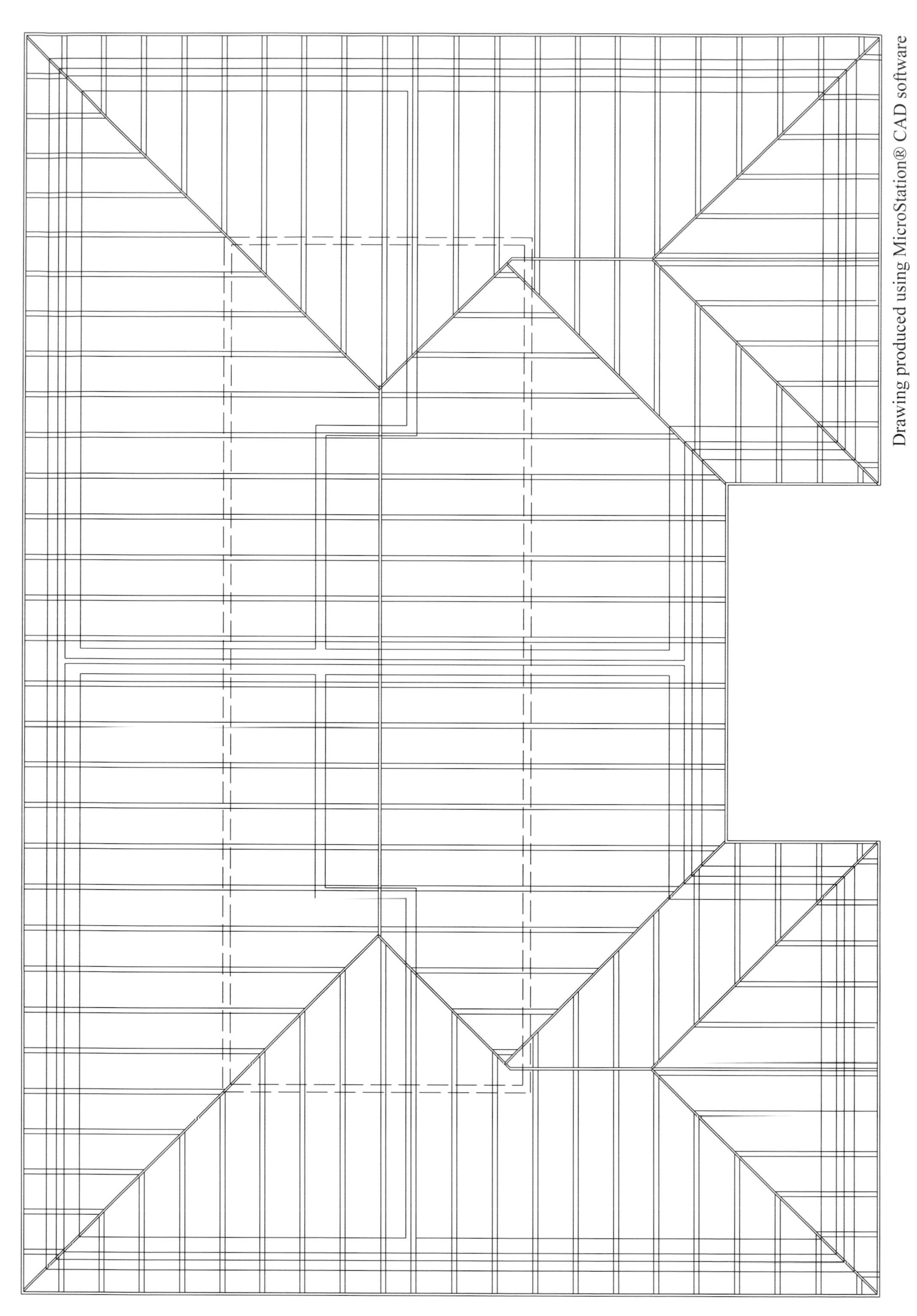

Andrew R Atkinson PhD MSc FRICS CertEd. Semi-Detached Houses. Drawing Q1/A4/2667-5C. Date 01-08-2023, Scale 1:50
ROOF CONSTRUCTION

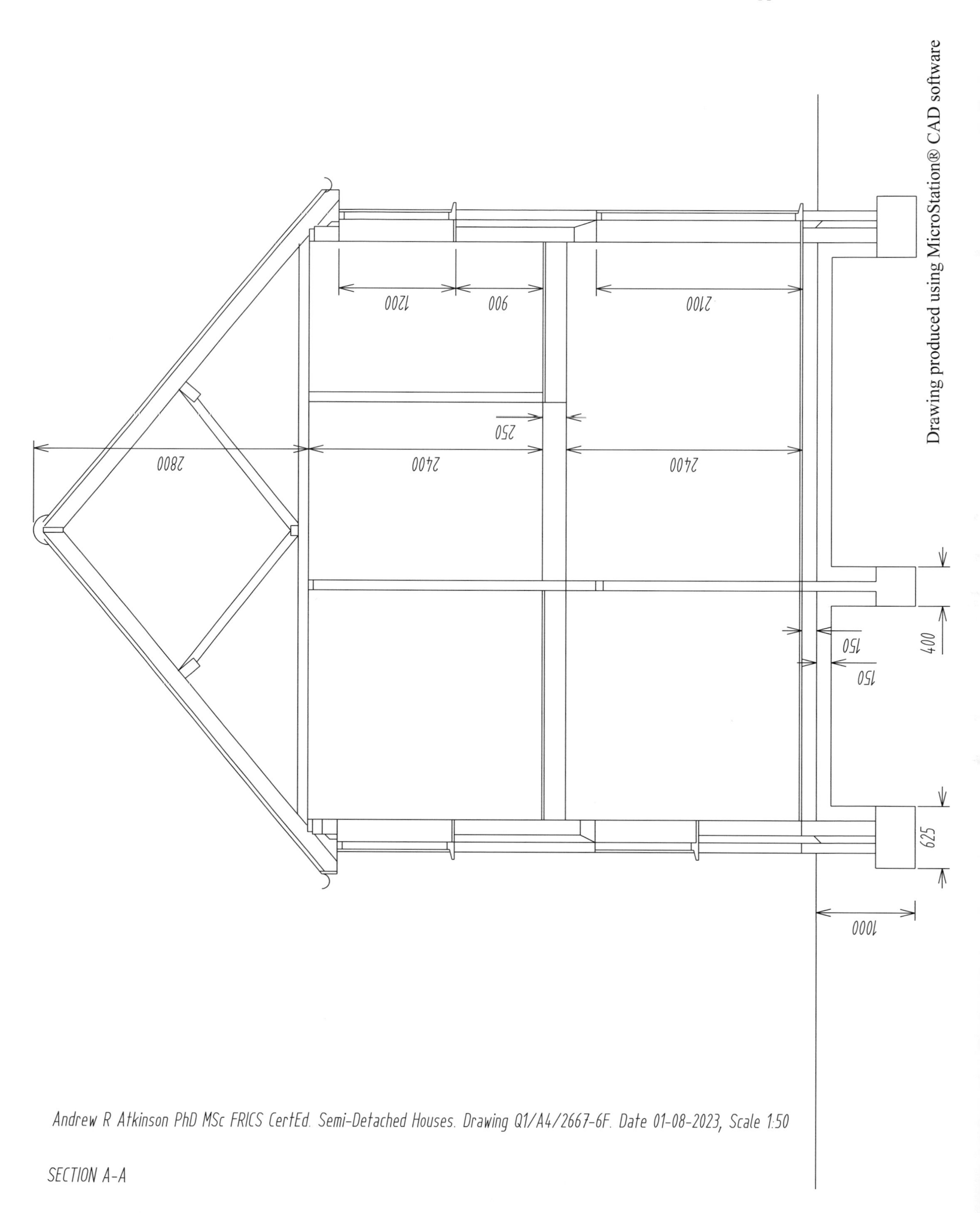

Drawing produced using MicroStation® CAD software

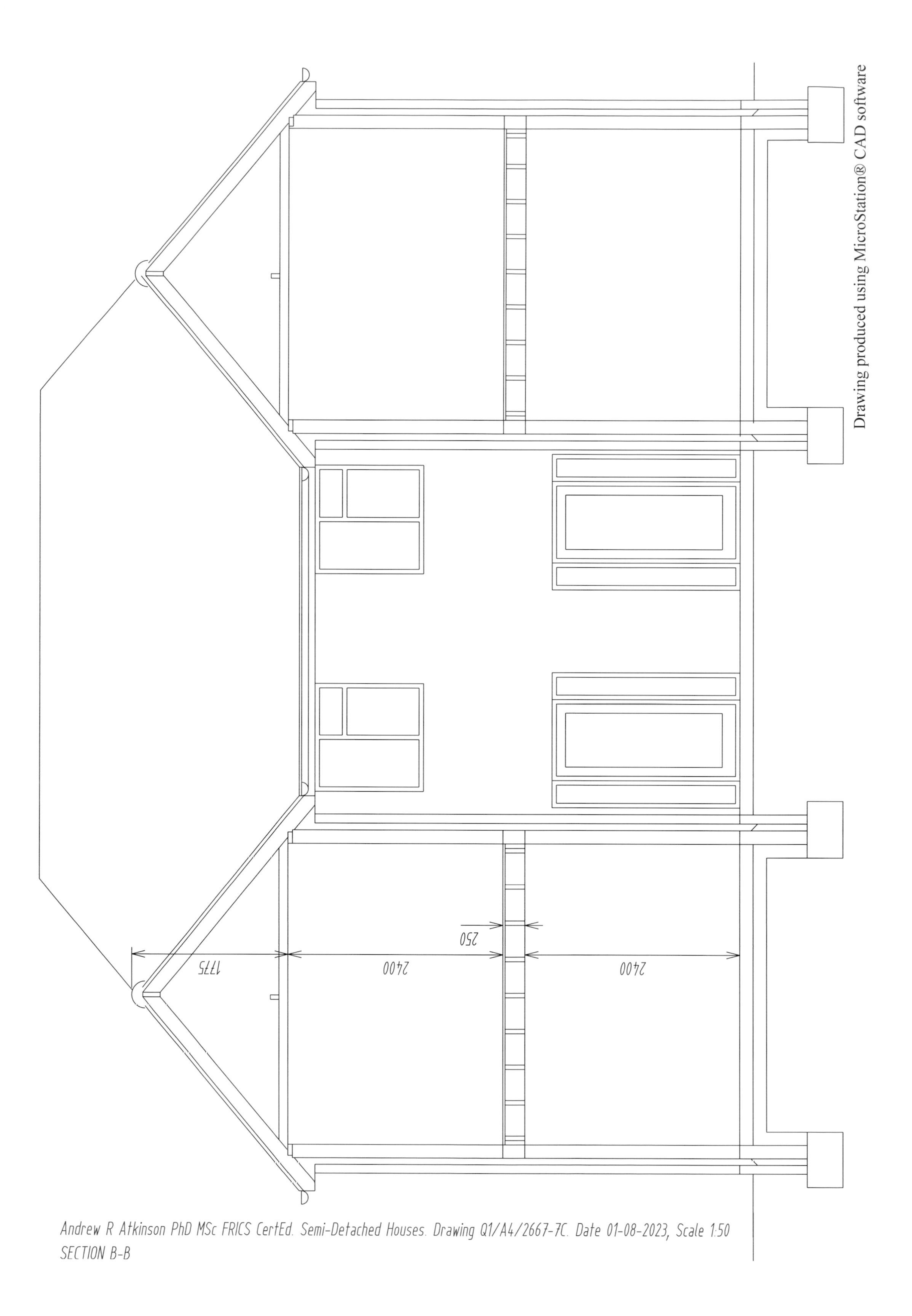

Andrew R Atkinson PhD MSc FRICS CertEd. Semi-Detached Houses. Drawing Q1/A4/2667-7C. Date 01-08-2023, Scale 1:50
SECTION B-B

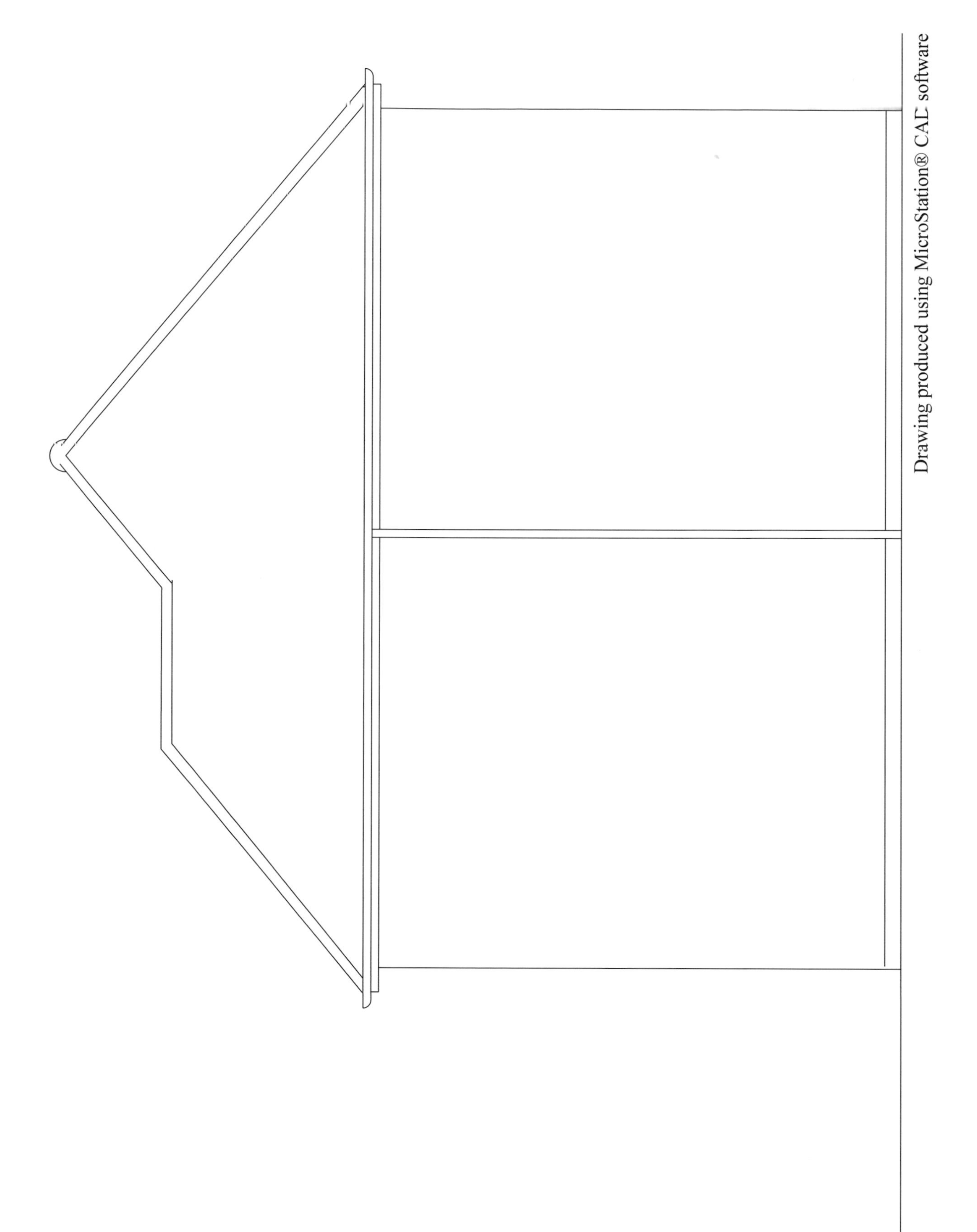

Andrew R Atkinson PhD MSc FRICS CertEd. Semi-Detached Houses. Drawing Q1/A4/2667-8. Date 01-08-2023, Scale 1:50
SIDE ELEVATION

Andrew R Atkinson PhD MSc FRICS CertEd. Semi-Detached Houses. Drawing Q1/A4/2667-9. Date 01-08-2023, Scale 1:50
FRONT ELEVATION

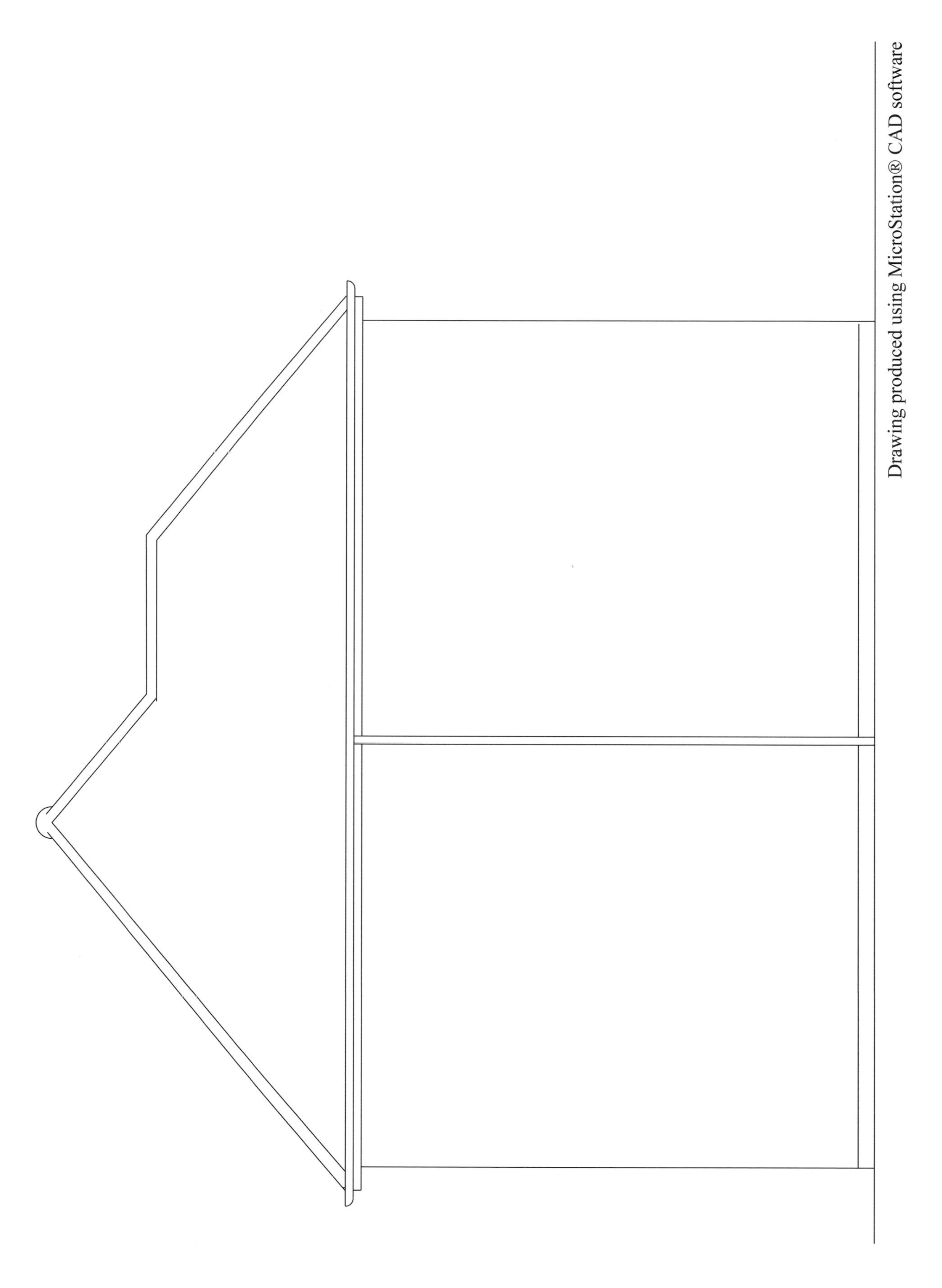

Andrew R Atkinson PhD MSc FRICS CertEd. Semi-Detached Houses. Drawing Q1/A4/2667-10 Date 01-08-2023, Scale 1:50
SIDE ELEVATION 2

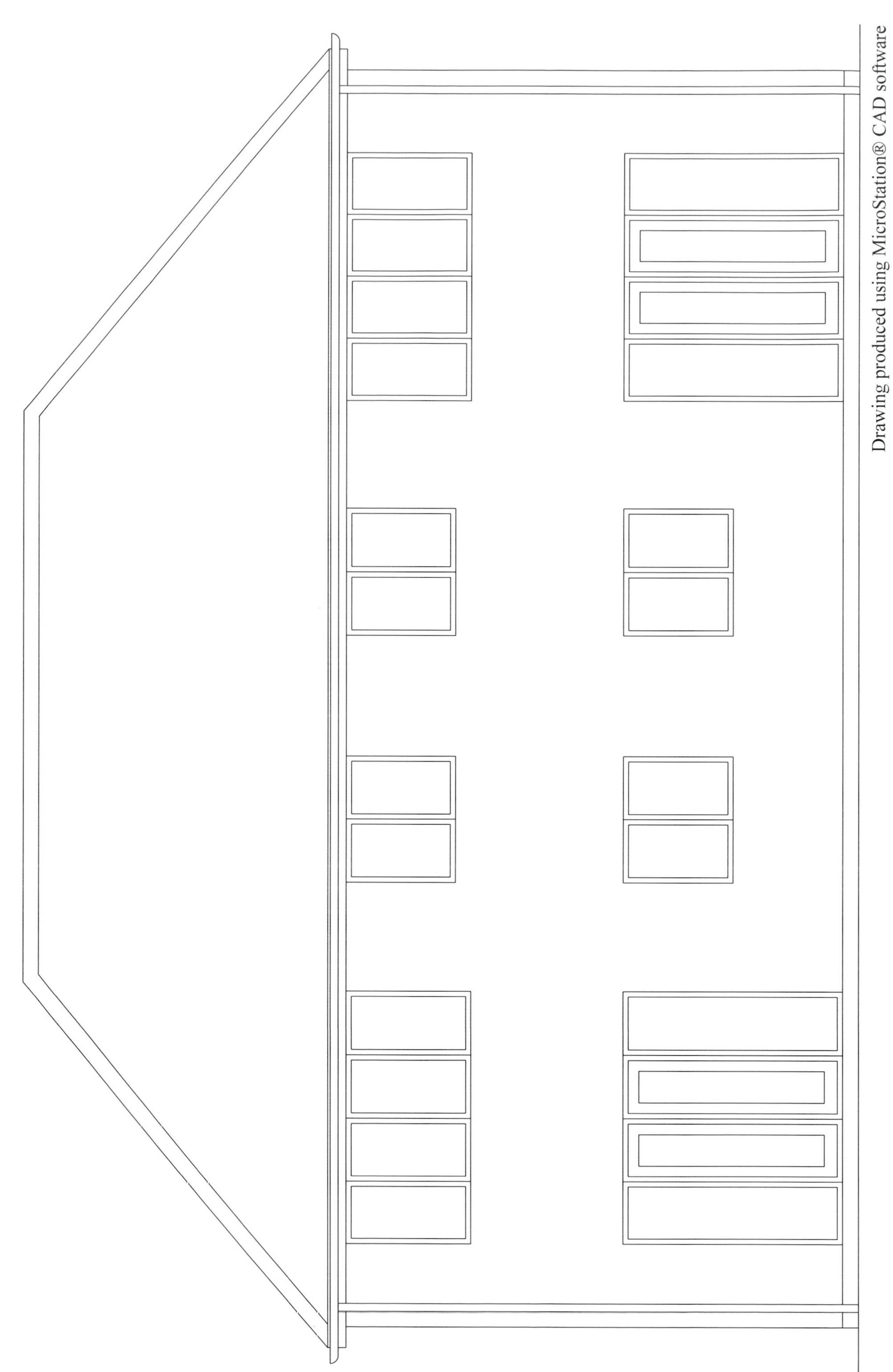

Andrew R Atkinson PhD MSc FRICS CertEd. Semi-Detached Houses. Drawing Q1/A4/2667-11 Date 01-08-2023, Scale 1:50
REAR ELEVATION

Appendix 4

Semi-detached houses brief specification

To be read in conjunction with detailed specification.

(Make any other assumptions necessary and add the assumptions to a query sheet.)

Substructure	
Topsoil	150mm vegetable soil to be excavated, retained, rotovated and spread and levelled on site 150mm thick, 20m from building.
Subsoil	To be removed from site.
Hardcore	Filling to trenches to be clean, hard, broken brick to pass a 50-mm screen well compacted in 200-mm layers.
	Oversite bed 150mm thick to be clean, hard, broken brick to pass a 50-mm screen blinded with sand to receive polythene damp-proof membrane.
Damp-proof membrane	2.7-micron polythene placed on blinded hardcore, lapped 300mm at joints and dressed up walls and into damp-proof course.
Concrete foundation	15N/mm^2 concrete with a 40-mm maximum aggregate trench foundation poured against earth.
Concrete bed	21N/mm^2 concrete with a 20-mm maximum aggregate 150mm thick poured over polythene damp-proof membrane.
Concrete cavity filling	21N/mm^2 concrete with a 10-mm maximum aggregate poured into 75-mm cavity (external walls) and 50-mm cavity (party wall) to within 150mm of damp-proof course.
Brickwork	Dense concrete (10.5N/mm^2) blocks laid in stretcher bond in cement mortar (1:4):
	100mm to outer skin of external walls, partitions and party wall skins. 150mm to inner skin of external walls.
	Allow four courses of facing bricks to outside face below damp-proof course – to be 102-mm Butterley Brown Brindle Bricks faced and pointed on one side with a neat weatherstruck joint in cement mortar (1:4).
	Form 75-mm cavity (external walls) and 50-mm cavity (party wall) with stainless steel wall ties spaced 600mm horizontally and vertically.
Damp-proof course	Single layer of vinyl lapped 150mm at joints.
Superstructure	
External walling	102-mm Butterley Brown Brindle Bricks faced and pointed on one side with a neat weatherstruck joint in gauged mortar (1:1:6).
	73-mm cavity with stainless steel wall ties spaced 600mm horizontally and vertically, fully filled with 73-mm glass-fibre resin treated wall bats. Continue insulation up to ceiling level behind inner skin of cavity walling.
	150-mm insulating concrete blocks in gauged mortar (1:1:6).

	Provide patent galvanised steel combined lintel and cavity tray over openings in cavity walls (Catnic CG70/125 or similar). Provide patent galvanised steel eaves lintel over openings under eaves (Catnic CGE90 or similar).
	Close cavity to openings at jambs with patent plastic combined cavity closer and damp-proof barrier. Leave cavity open at sills and eaves.
Internal walling	Loadbearing partitions to be 100-mm dense concrete blocks in gauged mortar (1:1:6). Non-loadbearing partitions to be 100-mm lightweight concrete blocks in gauged mortar (1:1:6). Party walls to be two skins of 100-mm dense concrete blocks in gauged mortar (1:1:6).
	Provide 65-mm-deep pre-stressed concrete lintels over internal openings in walls.
First floor	25-mm (nominal) tongued and grooved softwood flooring in 150-mm (nominal) face widths. Provide 40 × 19–mm finished nosing to exposed edge of stair opening. Boarding fixed to 225 × 50 softwood joists at 400-mm centres, built in. Provide single row of 38 × 38 double herringbone strutting to each span of joists.
	Provide two 225 × 50 joists nailed as trimmer and trimming joists to staircase opening and to support non-loadbearing partitions. Fix trimmer to trimming joists with 100 × 225 "Jiffy" galvanised joist hangar. Fix trimmed joists to trimmer joists with 50 × 225 "Jiffy" galvanised joist hanger.
Roof construction	150 × 50 rafters at 400-mm centres.
	100 × 50 ceiling joists at 400-mm centres.
	200 × 100 purlins at mid-span of rafters, to main slopes only.
	100 × 75 binder at mid-span of ceiling joists to main roof.
	100 × 50 binder at mid-span of ceiling joists to small roofs.
	100 × 50 plates at eaves and to loadbearing internal wall.
	200 × 32 ridge board.
	225 × 32 hip rafter.
	225 × 50 valley rafter.
	100 × 75 struts between purlin and binder – one to every fourth rafter, to main roof slopes only.
	Provide 250-mm glass-fibre insulation quilt at ceiling level.
Roof coverings	Plain concrete roof tiles to 40 degrees pitch and 100-mm gauge, nailed to 38 × 25 battens with aluminium nails. Provide layer of untearable and breathable vinyl under-tiling felt below battens.
	Half-round ridge tiles bedded in gauged mortar (1:1:6).
	Third-round hip tiles bedded in gauged mortar (1:1:6) with galvanised mild steel hip iron, painted with oil paint to end.
	Purpose-made valley tiles to valleys.
	Course of eaves tiles at eaves.
	Eaves fascia to be 150 × 25 (finished) softwood boarding once grooved for 15-mm exterior quality plywood soffit 200mm wide. Provide patent eaves ventilator over fascia and extending 450mm up roof slope.
	All exterior timber to be finished with traditional oil paint treatment.
Windows and external doors	See detailed specification.
	Allow a Defined Provisional Sum of £16000.00 for supply of window and external door units.

	All windows to be double glazed traditionally configured un-plasticised PVC to sizes as shown. To be delivered complete with ironmongery, glass, fixings and templates. Templates to be built into brickwork to give exact opening sizes for later fixing of window.
	All external doors to be double glazed traditionally configured un-plasticised PVC to sizes as shown. To be delivered complete with ironmongery, glass, fixings and templates. Templates to be built into brickwork to give exact opening sizes for later fixing of door
	Window boards to be supplied separately to windows, of 225 × 32–mm (finished) MDF with rounded front edge and tongue to rear edge, plugged and screwed to blockwork inner skin of wall. All interior timber to be finished with a traditional oil paint finish.
Internal doors and linings	Standard 38-mm plywood faced flush doors to suit blank opening fixed on two 100 × 100–mm pressed steel butts. Provide standard lever door handles and internal quality catch as detailed specification.
	125 × 32 (finished) softwood lining plugged and screwed to blockwork.
	Provide 50 × 16–mm (finished) softwood door stop planted on and 60 × 16–mm (finished) splayed and pencil rounded architrave. On closing and abutting door edges provide 25-mm (finished) quadrant in lieu of architrave.
	Loft hatch 750 × 750mm to be plywood faced flush panel as standard 38-mm door cut down on site, with 75 × 25–mm lining flush with ceiling finish fixed to softwood joists and 60 × 16–mm (finished) architrave all round. Trim joists for opening with standard ceiling joist on 50 × 100–mm "jiffy" pattern joist hangars.
Staircase	Allow a Defined Provisional Sum of £7500.00 for supply and fixing of two softwood staircases, open balustrading, trim and handrails complete by a specialist installer, to general configuration as shown on drawings.
	Provide 250 × 32–mm MDF lining to open edge of stairwell, with rounded bottom edge.
	Allow an Undefined Provisional Sum of £500.00 for additional builder's work in connection with the timber staircase.
Kitchen fittings	Allow a Defined Provisional Sum of £5000.00 for supply and fixing of standard kitchen fittings to two kitchens, to general configuration as shown on drawings.
Bathroom fittings	Allow a Defined Provisional Sum of £3000.00 for supply and fixing of standard bathroom fittings to two bathrooms, to general configuration as shown on drawings.
	Allow an Undefined Provisional Sum of £800.00 for builder's work in connection with kitchen and bathroom fittings.
	All interior timber to be finished with three coats of traditional oil paint.

Internal finishings schedule

Room	*Floors*	*Skirtings*	*Walls*	*Ceilings*	*Remarks*
Lounge/dining room	20-mm water-resistant MDF boarding laid as floating floor on 30-mm expanded polystyrene floor insulation.	150 × 25 ovalo mahogany stained and treated with two coats of polyurethane varnish, plugged, screwed and pelleted to blockwork.	Wallpaper Prime Cost Sum £60.00/piece. gypsum two-coat plaster of 10mm Browning rendering coat and 2.5mm Carlite finish setting coat.	12.5-mm plasterboard, skim coat plaster, three coats of emulsion. Provide 50 × 50–mm softwood noggings to end of plasterboard sheets to all ceilings.	150-mm Gyproc coving standard pattern to ceiling.
Hall	Ditto	150 × 25 ovalo softwood painted, plugged and screwed to blockwork.	Wallpaper Prime Cost Sum £40.00/piece. gypsum two-coat plaster.	Ditto	Ditto
Kitchen	Ditto	150 × 25 ovalo softwood painted.	Three coats of emulsion paint on gypsum two-coat plaster.	Ditto	Three courses of 150 × 150 × 4–mm wall tiles (Prime Cost Sum £40.00/m^2) over fittings. Coving as above.
Under-stairs cupboard	20-mm water-resistant MDF boarding laid as floating floor on 30-mm expanded polystyrene floor insulation.	150 × 25 ovalo softwood painted.	Three coats of emulsion paint on gypsum two-coat plaster.	12.5-mm plasterboard, skim coat plaster, three coats of emulsion to underside of staircase.	No coving.
Bedroom 1	150 × 25 softwood tongued and grooved floor boarding.	Ditto	Ditto	12.5-mm plasterboard, skim coat plaster, three coats of emulsion. Provide 50 × 50–mm softwood noggings to end of plasterboard sheets to all ceilings.	Ditto
Bedroom 2	Ditto	Ditto	Ditto	Ditto	Ditto
Bedroom 3	Ditto	Ditto	Ditto	Ditto	Ditto
Bathroom	Ditto	Ditto	150 × 150 × 4–mm wall tiles (Prime Cost Sum £30.00/m2) on gypsum two-coat plaster.	Ditto	Ditto

Appendix 5

The exercise specification

Introduction

Appendices 5 and 6 contain the practice exercise. Taking off involves most of the same sections as in the examples in the book (except that there is no upper floor!) and the work is similar but sufficiently different to make you think. This is the best way to learn measurement – you can replicate most of the items, but the physical measurements and specification vary slightly. Much of the work is straightforward, designed to give confidence that you are progressing.

In tackling the exercise, treat each section separately and don't be tempted to take on too much too quickly. It is best to set the example taking off alongside your work, with the drawings for both example and exercise easily to hand. In learning measurement, it is often beneficial (though not essential) to work cooperatively. If you were working as an apprentice in an office, you would be given support and guidance and be able to ask co-workers how things are done. So, unless you are being formally assessed, it is more efficient and accurate if you work with others. Even in exam situations, treating a taking-off test as a "timed assignment", allowing full use of office resources, support and advice, is more beneficial than working in isolation. Taking off in theory and practice is not a memory test but is the development of sound method.

Before starting a section of taking off, remember the preliminary tasks of preparing drawings, checking dimensions, drafting query sheets, producing taking-off lists and so on. These are designed to give an overview of the construction and to allow you to plan your approach to the work. Executing these tasks will make the actual measurement quicker and easier.

There follows the brief specification for the exercise – a typical traditional single-storey "bungalow" as seen throughout the suburbs in the United Kingdom. Appendix 6 shows the drawings set out in a similar way to Appendix 3. I hope you enjoy tackling the exercise; on completion you should be on your way to technical competence in the specialism.

Bungalow 2023 brief specification

To be read in conjunction with detailed specification (not supplied).

(Make any other assumptions necessary and add the assumptions to a query sheet.)

Substructure	
Topsoil	150mm vegetable soil to be excavated, retained and spread and levelled on site 150mm thick, 25m from building.
Subsoil	To be removed from site.
Hardcore	Filling to trenches to be clean, hard, broken brick to pass a 50-mm screen well compacted in 200-mm layers.
	Oversite bed 150mm thick to be clean, hard, broken brick to pass a 50-mm screen blinded with sand to receive polythene damp-proof membrane.

Damp-proof membrane	2.7-micron polythene laid on hardcore, lapped 300mm at joints and dressed up walls and into damp-proof course.
Concrete foundation	15N/mm^2 concrete with a 40-mm maximum aggregate foundation poured against earth.
Concrete bed	21N/mm^2 concrete with a 20-mm maximum aggregate 150mm thick poured over polythene damp-proof membrane.
Concrete cavity filling	21N/mm^2 concrete with a 10-mm maximum aggregate poured into 75-mm cavity to within 150mm of damp-proof course.
Brickwork	Dense concrete (10.5N/mm^2) blocks laid in stretcher bond in cement mortar (1:4):
	Allow four courses of facing bricks to outside face below damp-proof course – to be 102-mm Butterley Brown Brindle Bricks faced and pointed on one side with a neat weatherstruck joint in cement mortar (1:4).
	Form 75-mm cavity with stainless steel wall ties spaced 450mm horizontally and vertically.
Damp-proof course	Single layer of vinyl lapped 150mm at joints.
Superstructure	
External walling	102-mm Butterley Brown Brindle Bricks faced and pointed on one side with a neat weatherstruck joint in gauged mortar (1:1:6).
	73-mm cavity with stainless steel wall ties spaced 450mm horizontally and vertically, fully filled with 73-mm glass-fibre resin-treated wall bats. Continue insulation up to ceiling level behind inner skin of cavity walling.
	100-mm insulating concrete blocks in gauged mortar (1:1:6).
	Provide patent galvanised steel combined lintel and cavity tray over openings in cavity wall gables (Catnic CG70/125 or similar). Provide patent galvanised steel eaves lintel over openings under eaves (Catnic CGE90 or similar).
	Close cavity to openings at jambs with patent plastic combined cavity closer and damp-proof barrier. Leave cavity open at sills and eaves.
Internal walling	Loadbearing and non-loadbearing partitions to be 100-mm dense concrete blocks in gauged mortar (1:1:6).
	Provide 65-mm-deep pre-stressed concrete lintels over internal openings in walls.
Roof construction	150 × 50 rafters at 400-mm centres.
	150 × 50 ceiling joists at 400-mm centres.
	175 × 100 purlins at mid-span of rafters.
	100 × 75 binder at mid-span of ceiling joists.
	100 × 75 struts between purlin and binder as shown on drawing.
	100 × 50 hangar between ridge and binder as shown on drawing.
	100 × 50 plates at eaves and to loadbearing internal wall.
	225 × 25 ridge board.
	225 × 32 hip rafter.
	Provide 250mm glass-fibre insulation at ceiling level.
Roof coverings	Patent interlocking concrete roof tiles to 30 degrees pitch and 300mm gauge, nailed to 38 × 25 battens with aluminium nails. Provide layer of un-tearable and breathable vinyl under-tiling felt below battens.
	Half-round ridge tiles bedded in gauged mortar (1:1:6). Third-round hip tiles bedded in gauged mortar (1:1:6).
	Eaves fascia to be 200 × 25 (finished) softwood boarding once grooved for 15-mm exterior quality plywood soffit 275mm wide. Provide patent plastic eaves ventilator over fascia and extending 450mm up roof slope.
	All exterior timber to be finished with traditional oil paint treatment.

Windows and doors	See detailed specification. Allow a Defined Provisional Sum of £12000.00 for supply of window and external door units. All windows to be double-glazed traditionally configured un-plasticised PVC to sizes as shown. To be delivered complete with ironmongery, glass, fixings and templates. Templates to be built into brickwork to give exact opening sizes for later fixing of window. All external doors to be double-glazed traditionally configured un-plasticised PVC to sizes as shown. To be delivered complete with ironmongery, glass, fixings and templates. Templates to be built into brickwork to give exact opening sizes for later fixing of door. Window boards to be supplied separately to windows, of 175 × 25–mm (finished) softwood with rounded front edge and tongue to rear edge, plugged and screwed to blockwork inner skin of wall.
Internal doors and linings	Standard 38-mm plywood faced flush doors to suit blank opening fixed on two 100 × 100–mm pressed steel butts. Allow the Prime Cost Sum of £50.00 per set for the supply of internal door ironmongery. Fix standard lever door handles and internal quality catch as detailed specification. 125 × 32 (finished) softwood lining plugged and screwed to blockwork. Provide 50 × 16–mm (finished) softwood door stop planted on and 60 × 16–mm (finished) splayed and pencil rounded architrave. On closing and abutting door edges provide 25-mm (finished) quadrant in lieu of architrave. Loft hatch 750 × 750mm to be plywood faced flush panel as standard 38-mm door cut down on site, with 75 × 25–mm lining flush with ceiling finish fixed to softwood joists and 60 × 16–mm (finished) architrave all round. Trim joists for opening with standard ceiling joist on 50 × 100–mm "jiffy" pattern joist hangars.
Kitchen fittings	Allow a Defined Provisional Sum of £4000.00 for supply and fixing of standard kitchen fittings to two kitchens, to general configuration as shown on drawings.
Bathroom fittings	Allow a Defined Provisional Sum of £2000.00 for supply and fixing of standard bathroom fittings to two bathrooms, to general configuration as shown on drawings. Allow an Undefined Provisional Sum of £800.00 for builder's work in connection with kitchen and bathroom fittings. All interior timber to be finished with three coats of traditional oil paint.

Internal finishings schedule

Room	*Floors*	*Skirtings*	*Walls*	*Ceilings*	*Remarks*
Sitting room/dining room	20-mm water-resistant MDF boarding laid as floating floor on 30-mm expanded polystyrene floor insulation.	150 × 25 ovalo oak stained and treated with two coats of polyurethane varnish, plugged, screwed and pelleted to blockwork.	Wallpaper Prime Cost Sum £40.00/piece. gypsum two-coat plaster of 10-mm Browning rendering coat and 2.5-mm Carlite finish setting coat.	12.5-mm plasterboard, skim coat plaster, three coats of emulsion. Provide 50 × 50–mm softwood noggings to end of plasterboard sheets to all ceilings.	150-mm Gyproc coving standard pattern to ceiling.
Hall	Ditto	150 × 25 ovalo softwood painted plugged and screwed to blockwork.	Wallpaper Prime Cost Sum £30.00/piece. gypsum two-coat plaster	Ditto	Ditto
Kitchen	Ditto	150 × 25 ovalo softwood painted.	Three coats of emulsion paint on gypsum two-coat plaster.	Ditto	Four courses of 150 × 150 × 4–mm wall tiles (Prime Cost sum £35.00/m^2) over fittings. Coving as above.
Bedroom 1	Ditto	Ditto	Three coats of emulsion paint on gypsum two-coat plaster.	Ditto	No coving.
Bedroom 2	Ditto	Ditto	Ditto	Ditto	Ditto
Bathroom	Ditto	Ditto	150 × 150 × 4–mm wall tiles (Prime Cost sum £35.00/m^2) on gypsum two-coat plaster.	Ditto	Ditto
Airing cupboard	Ditto	Ditto	Three coats of emulsion paint on gypsum two-coat plaster.	Ditto	Ditto

Appendix 6

The exercise drawings

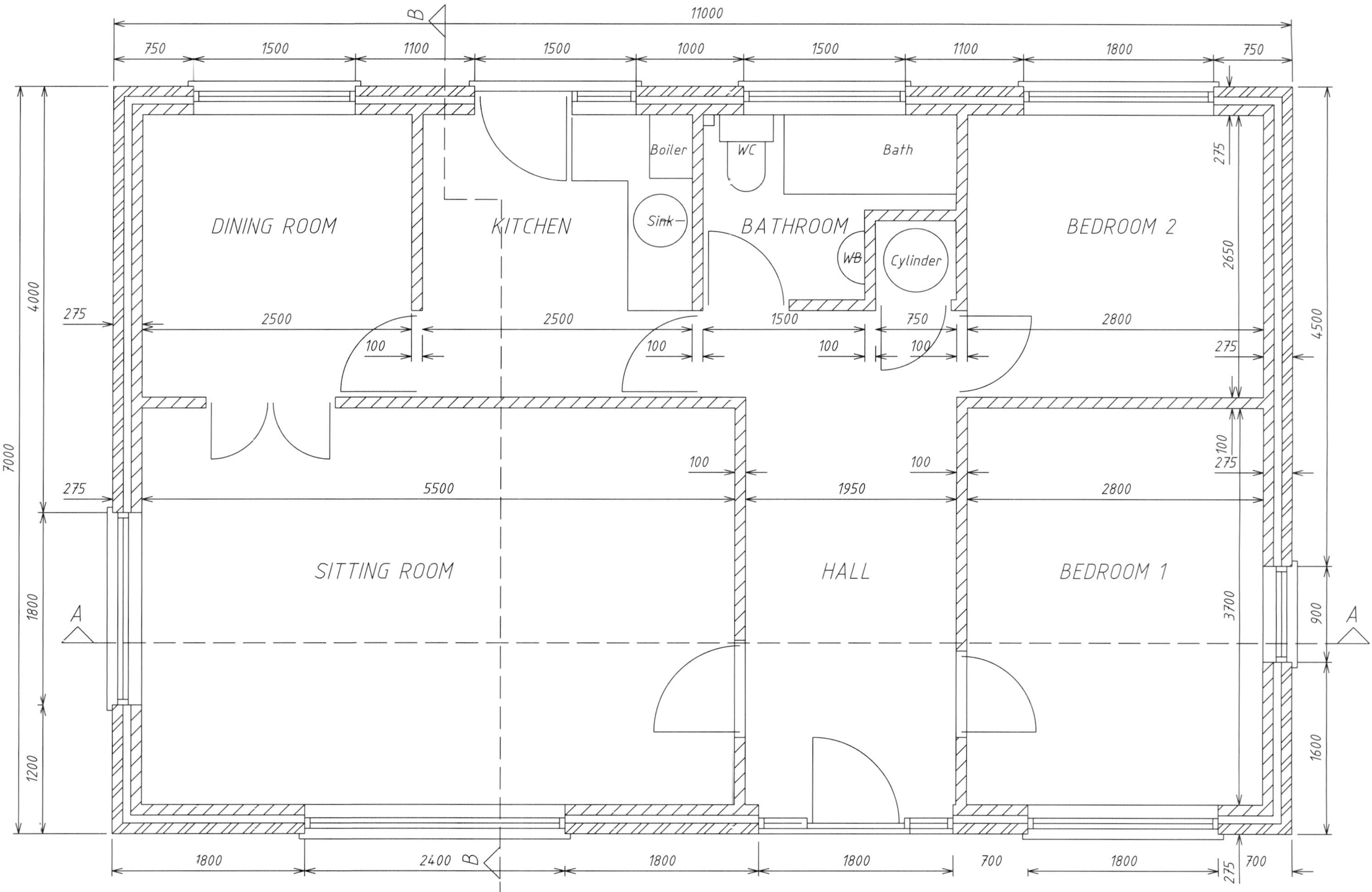

DRAWING NO Q1/A4/9595/1a DATE 20-08-2023, SCALE 1:50
DRAWN BY ANDREW R ATKINSON PhD MSc FRICS
NEW BUNGALOW FLOOR PLAN

Drawing produced using MicroStation® CAD software

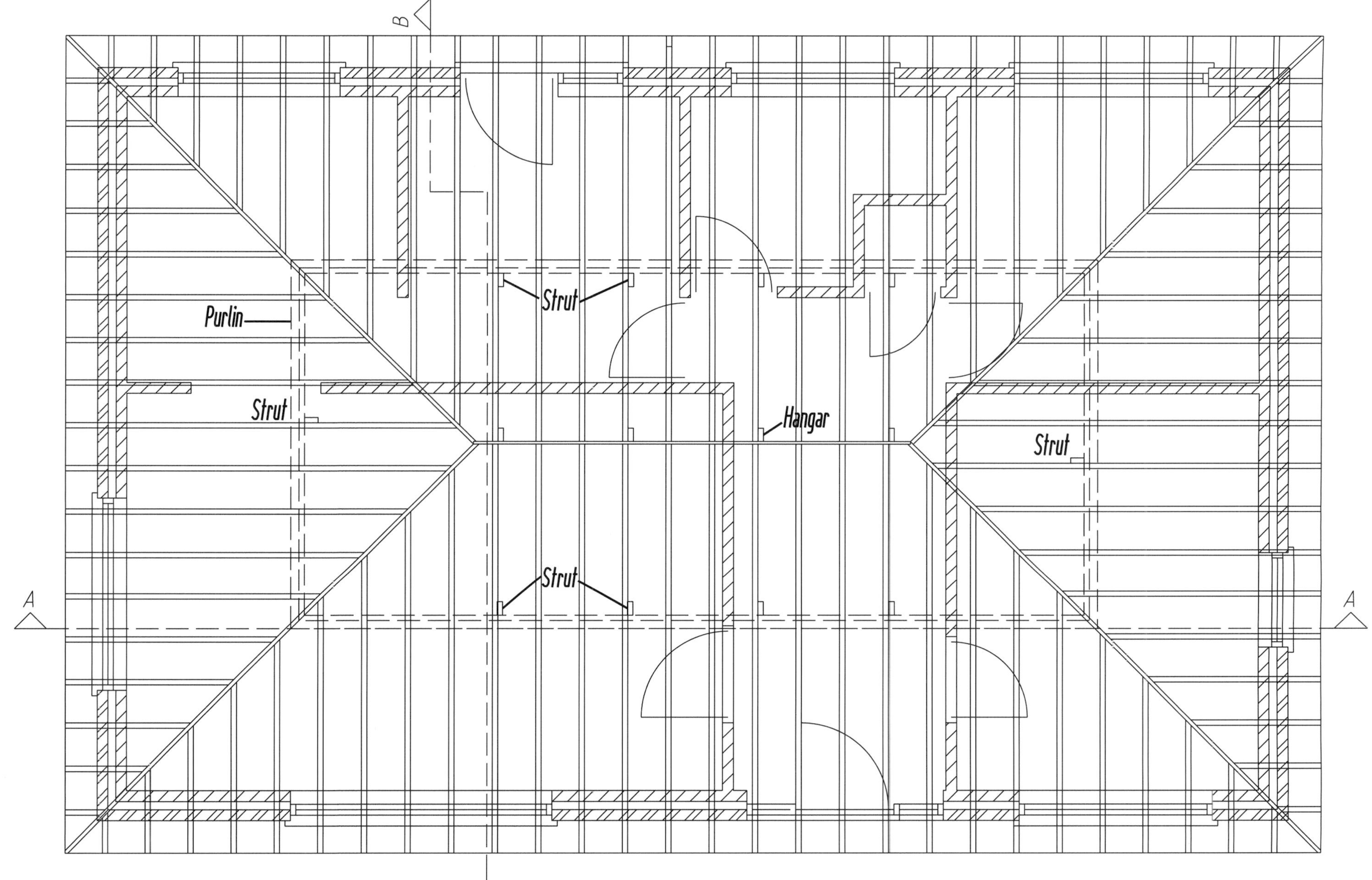

DRAWING NO Q1/A4/9595/8c DATE 20-08-2023, SCALE 1:50
DRAWN BY ANDREW R ATKINSON PhD MSc FRICS
NEW BUNGALOW ROOF PLAN

Drawing produced using MicroStation® CAD software

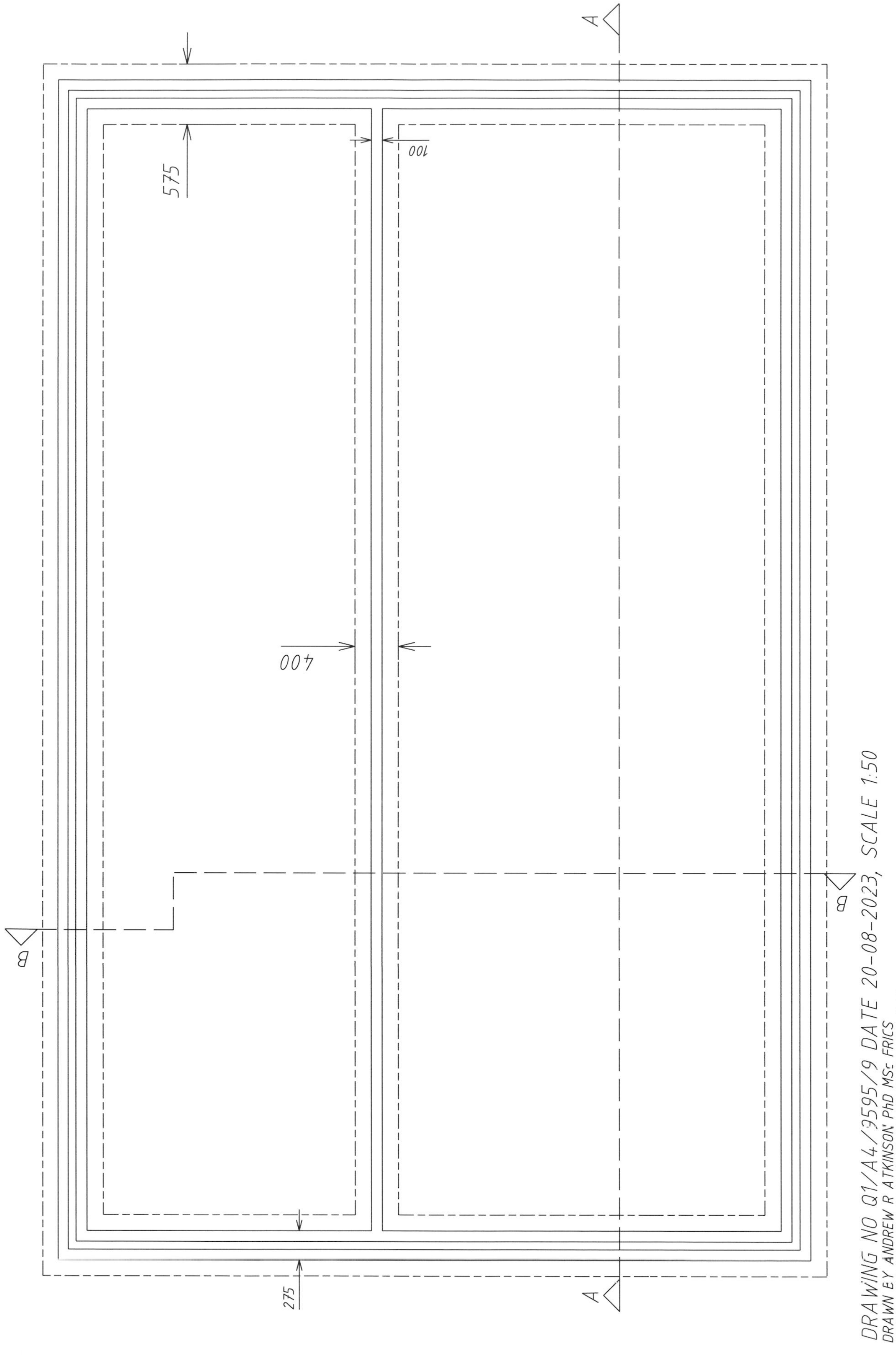

A
A
B
B
575
100
400
275

DRAWING NO Q1/A4/3595/9 DATE 20-08-2023, SCALE 1:50
DRAWN BY ANDREW R ATKINSON PhD MSc FRICS
NEW BUNGALOW FOUNDATION PLAN

Drawing produced using MicroStation® CAD software

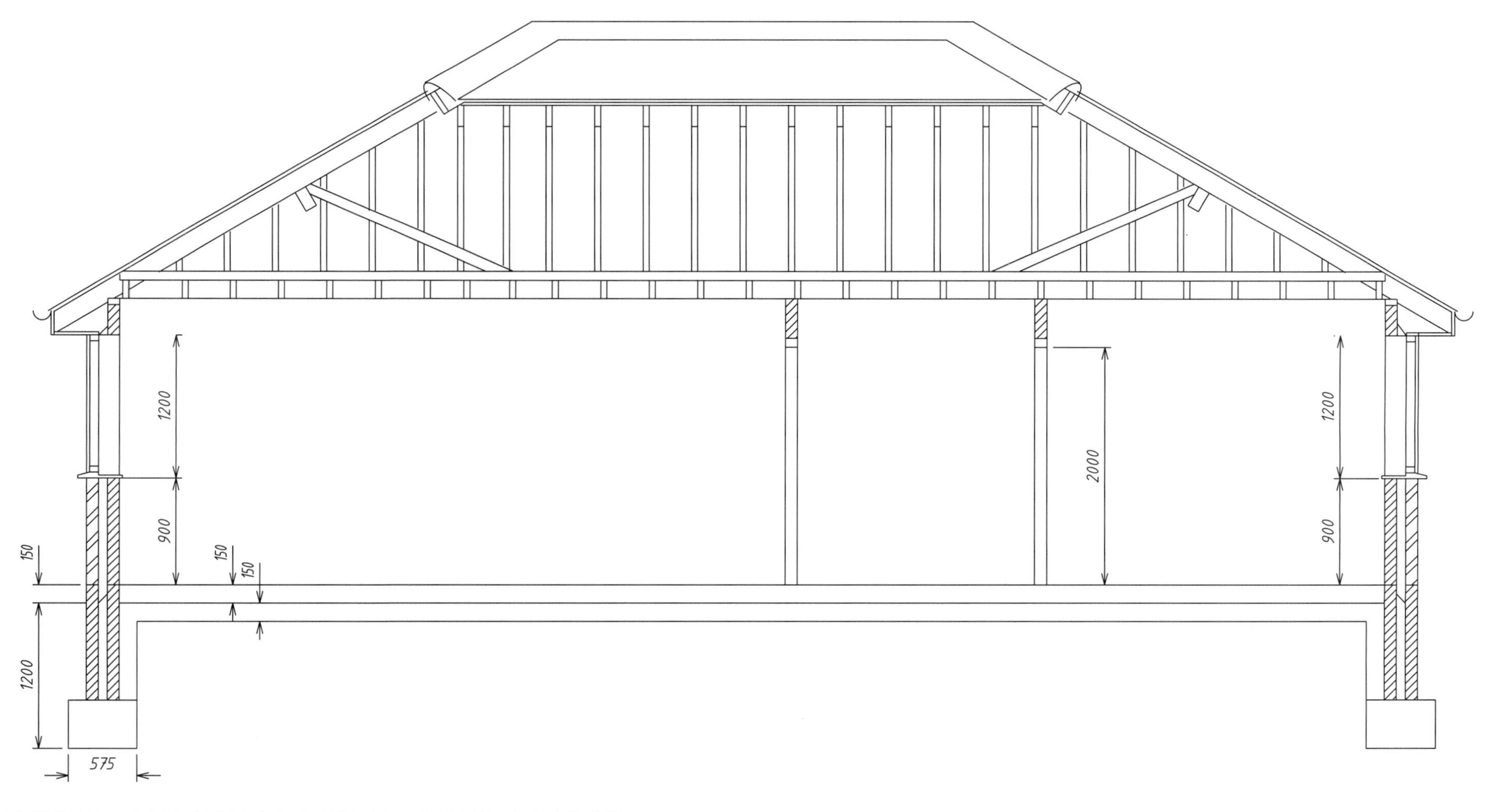

Drawing produced using MicroStation® CAD software

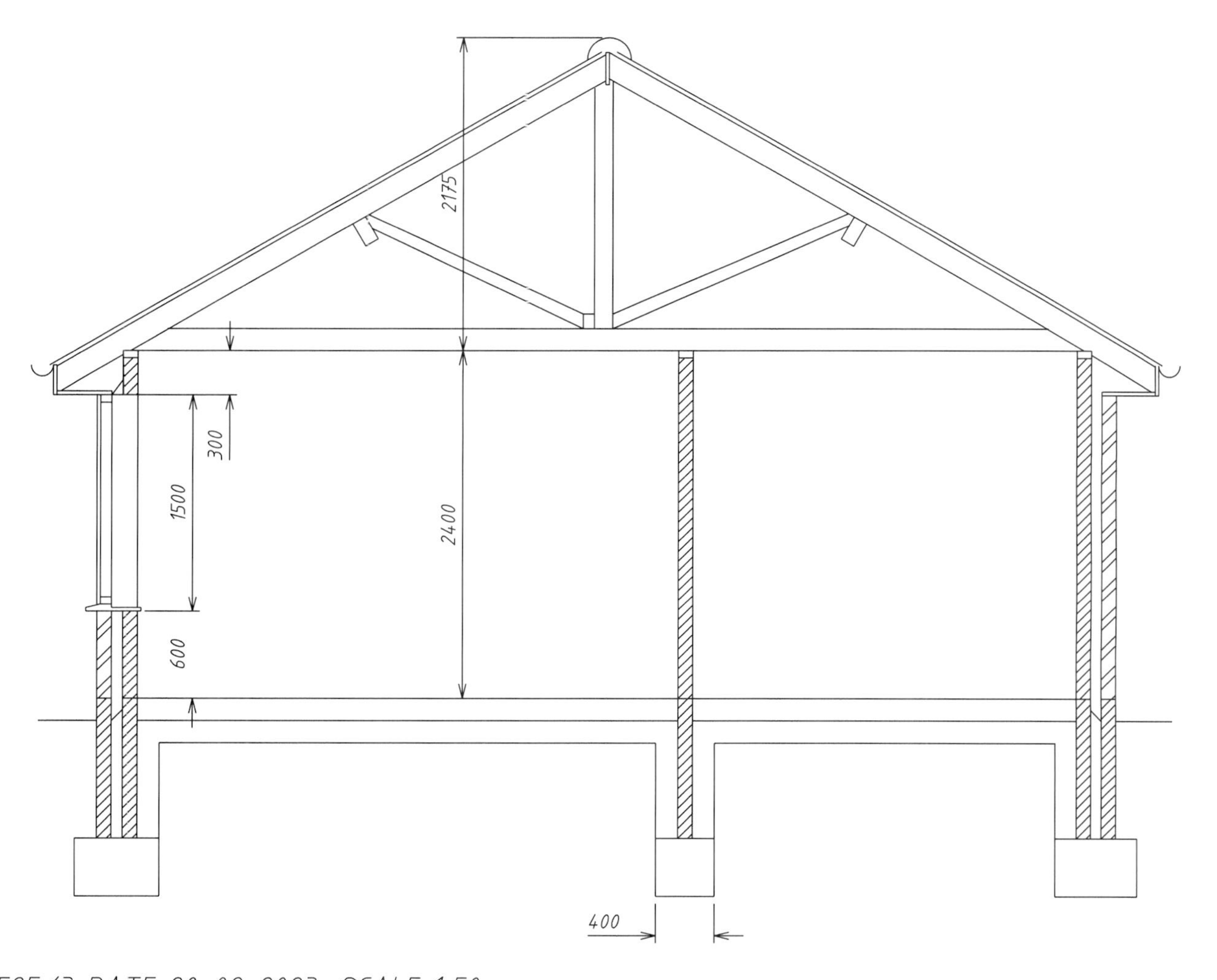

DRAWING NO Q1/A4/9595/3 DATE 20-08-2023, SCALE 1:50,
DRAWN BY ANDREW R ATKINSON PhD MSc FRICS
NEW BUNGALOW SECTION B-B

Drawing produced using MicroStation® CAD software

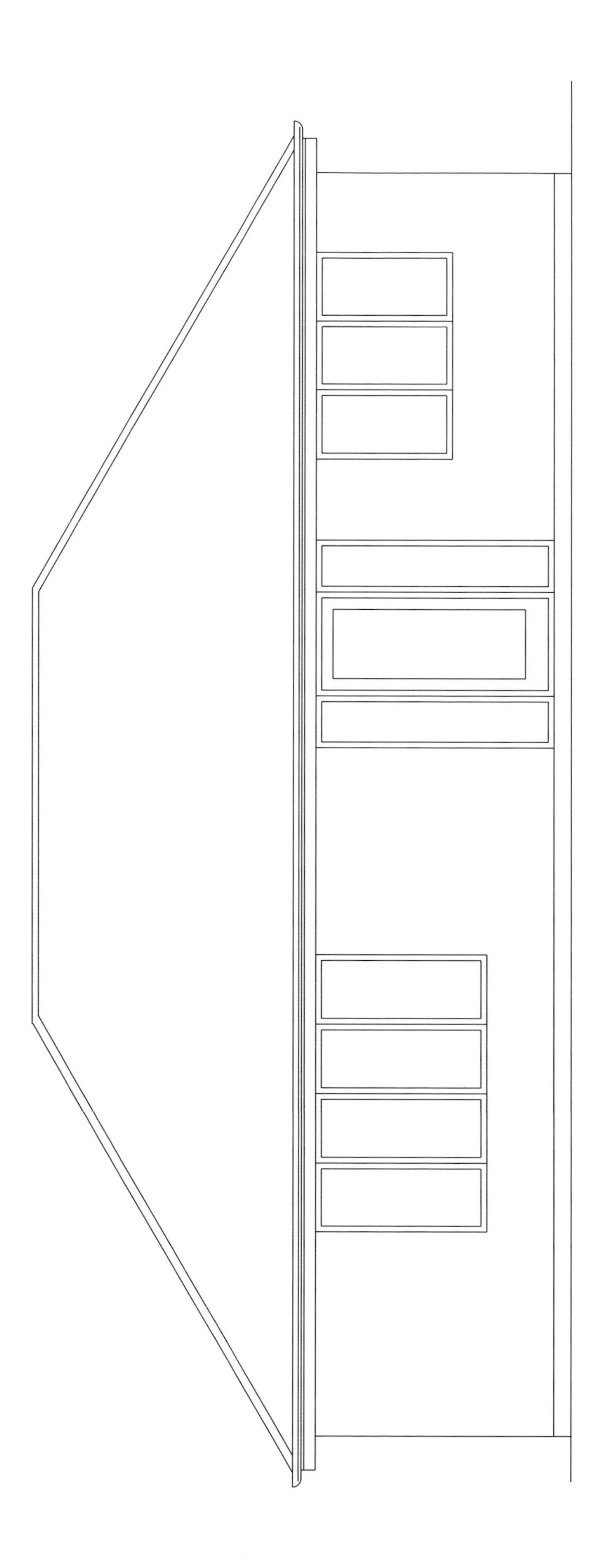

Drawing produced using MicroStation® CAD software

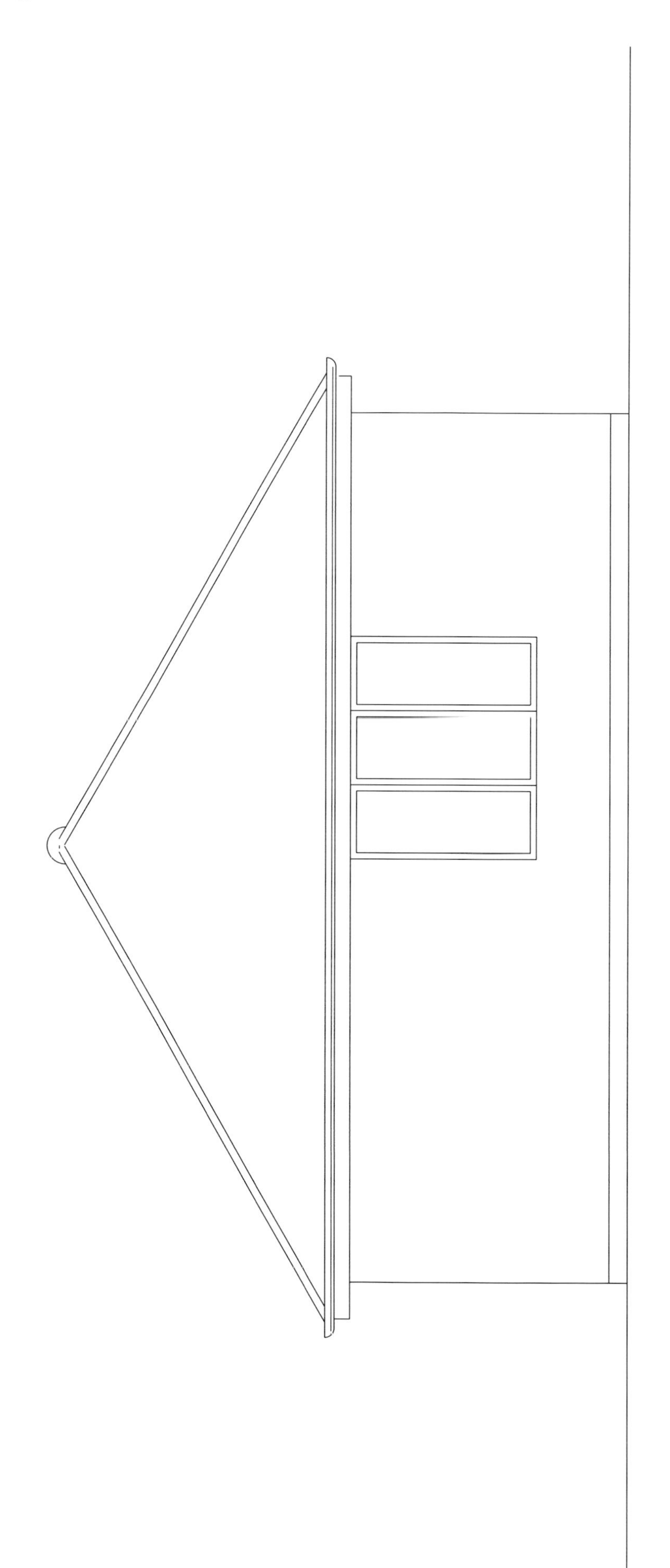

DRAWING NC Q1/A4/9595/6 DATE 20-08-2023, SCALE 1:50,
DRAWN BY ANDREW R ATKINSON PhD MSc FRICS
NEW BUNGALOW SIDE ELEVATION 1

Drawing produced using MicroStation® CAD software

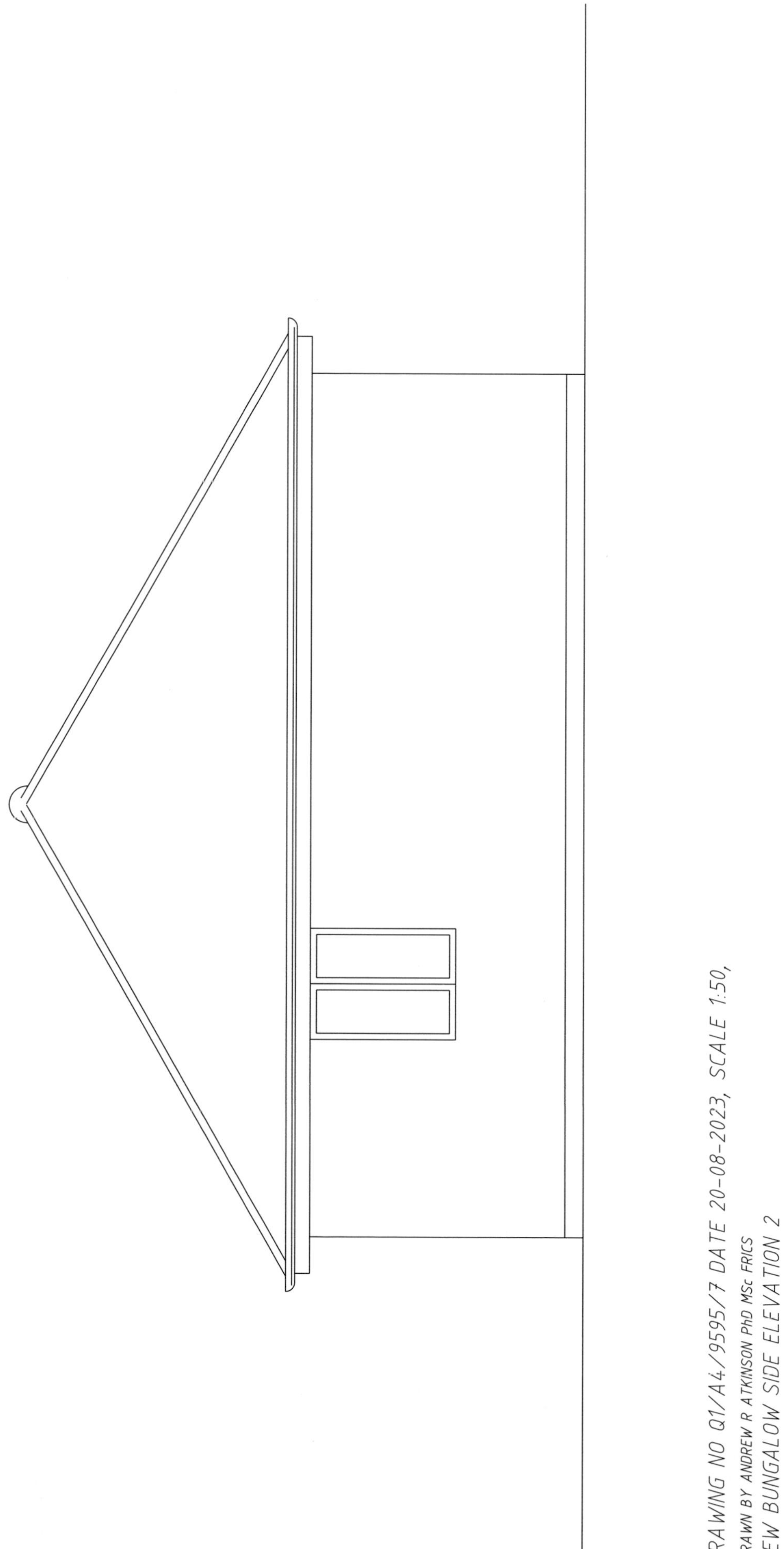

DRAWING NO Q1/A4/9595/7 DATE 20-08-2023, SCALE 1:50,
DRAWN BY ANDREW R ATKINSON PhD MSc FRICS
NEW BUNGALOW SIDE ELEVATION 2

Drawing produced using MicroStation® CAD software

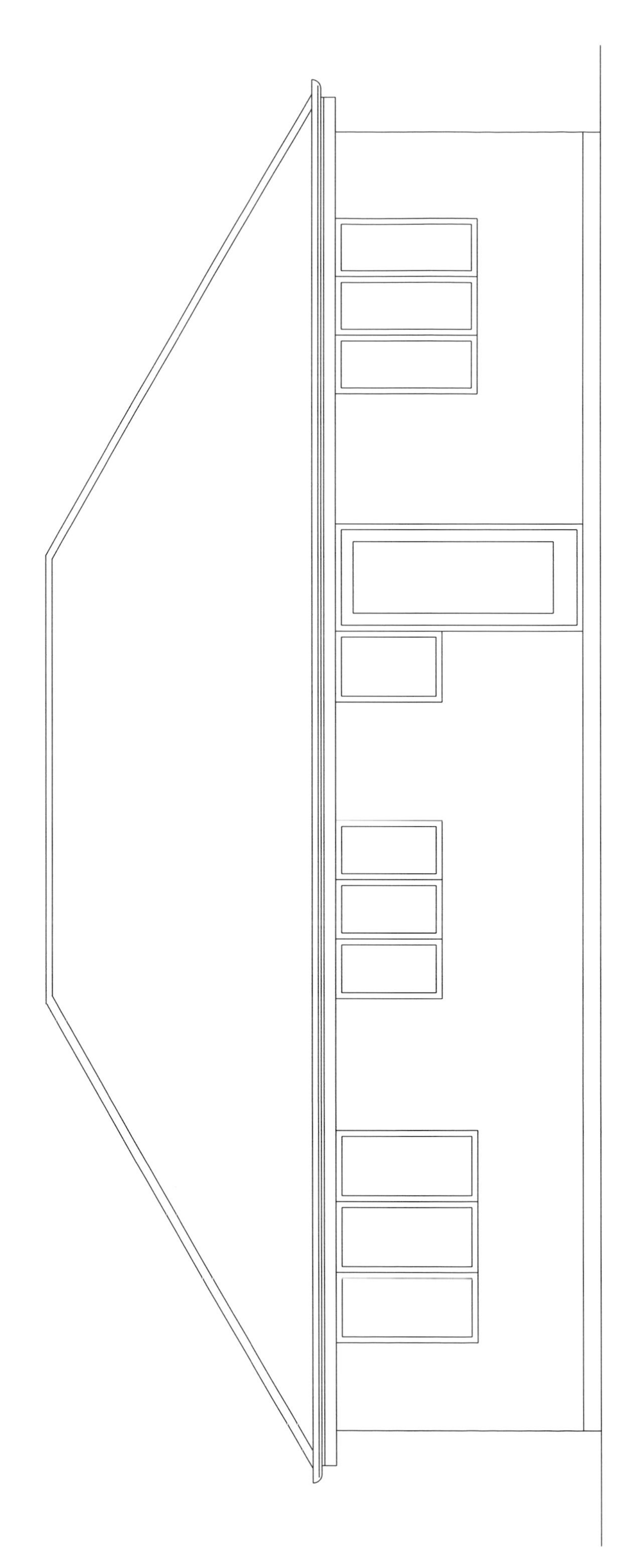

Drawing produced using MicroStation® CAD software

Index